Gender's Place

Gender's Place

Feminist Anthropologies of Latin America

Edited by

*Rosario Montoya, Lessie Jo Frazier, and
Janise Hurtig*

Rosa Acle (b. 1916, Rio de Janeiro, Brazil, d. 1990, Montevidoo, Uruguay) studied with the Uruguayan painter Joaquín Torros-García (1874–1949) [the School of the South] and became a member of the Asociación de Arte Constructivo which reproduced her works in its magazine "Círculo y Cuadrado." Her first solo exposition was in the Montevideo "Amigos de Arte" (1939) after which she traveled to Europe and Australia, remaining in Melbourne until returning to Uruguay 1947. At the time of her death, she was preparing what became her posthumous retrospective exposition in the Galería Montevideo, Uruguay. This painting, Norte, is an architectonic cultural mapping of the Americas juxtaposing indigenist and modernist symbols to invert dominant cartographies of power in the region.

First published 2002 by PALGRAVE MACMILLAN™
175 Fifth Avenue, New York, N.Y. 10010 and
Houndmills, Basingstoke, Hampshire, England RG21 6XS.
Companies and representatives throughout the world.

PALGRAVE MACMILLAN is the global academic imprint of the Palgrave Macmillan division of St. Martin's Press, LLC and of Palgrave Macmillan Ltd. Macmillan® is a registered trademark in the United States, United Kingdom and other countries. Palgrave is a registered trademark in the European Union and other countries.

ISBN 1–4039–6039–9 (hardback)   1–4039–6040–2 (paperback)

Library of Congress Cataloging-in-Publication Data
Gender's Place: feminist anthropologies of Latin America/edited by Rosario Montoya, Lessie Jo Frazier, and Janise Hurtig.
     p. cm.
     Includes bibliographical references and index.
     ISBN 1–4039–6039–9 (hardback) 1–4039–6040–2 (paperback)
     1. Sex role—Latin America. 2. Women—Latin America—Social conditions.
3. Feminist anthropology—Latin America. 4. Feminist theory—Latin America.
I Montoya, Rosario, 1960– II. Frazier, Lessie Jo, 1966– III. Hurtig, Janise.

HQ1075.5L29 G462 2002
305.42'098—dc21   2002074942

A catalogue record for this book is available from the British Library.

Design by Newgen Imaging Systems (P) Ltd., Chennai, India.

First edition: November, 2002
10 9 8 7 6 5 4 3 2 1

Printed in the United States of America.
Transferred to Digital Printing 2005

Contents ✍

Acknowledgements ↶

The editors would like to thank our Palgrave editor, Kristi Long, who has offered enthusiastic support and critical feedback for this project, patiently guiding us through the publication process. For their insightful comments on the entire manuscript, we would like to thank Susan Paulson and Yvonne Reineke. We also thank Ileana Rodríguez for her creative and constructive response to the Latin American Studies Association panel out of which this volume grew, and for her ongoing encouragement. For feedback on several of the papers, we are grateful to participants in the History/Latin American Studies Gender Workshop, University of Illinois at Chicago, directed by Mary Kay Vaughn (April 1995). Partial funding for this project was provided by the University of Michigan Institute for Research on Women and Gender. This book could not have been completed without the technical support in editing, translating, formatting, and indexing we received from Tina Meltzer, Deborah Cohen, and Jodi Barns as well as the editorial assistance from Sonia Wilson, Roee Raz, and Annje Kern at Palgrave Macmillan. Finally, we would like to acknowledge the inspiration Daniel Viglietti's music and especially his song "A Desalambrar" contributed to this work. We thank Daniel for graciously allowing us to reproduce portions of the text in our introduction.

Preface ᔮ

Gender Que Pica un Poco

Ruth Behar

Now, at the dawn of our new century, it's difficult to imagine a time when anthropologists didn't have the concept of gender. But it really wasn't all that long ago that we traveled around the world, seeing gender everywhere but not really seeing it because we didn't have a name for what we saw. Back then, we didn't even know that the anthropologist, no less than the informant, was also gendered!

How quickly things can change. How quickly intellectual self-awareness can take root. These days an anthropological account that is totally oblivious to gender is unthinkable. It would be possible to write a history of anthropology based entirely on the rupture between the anthropology that was done before and after the concept of gender was adopted. Ever since anthropologists acquired the concept of gender following the feminist awakening of the 1970s, we have used it so widely and so rampantly that we are close to reaching a kind of paradigm fatigue about gender. What more can be said about this most basic and most necessary of analytical perspectives? Is gender so self-evident to us at this point that we have, ironically, come around full circle to our original blindness toward gender? Why should we care about gender anymore?

The volume before us argues that gender still has a central place in our research and writing, that we have not yet exhausted all of the possibilities for thinking through gender in anthropology. Its singular contribution is its eclectic willingness to combine symbolic and material approaches to gender, to see these two realms of human existence as interconnected. This volume further argues that Latin America offers an especially exciting location both for reflecting on the place of gender and for doing gender as a form of critical practice. Seeing gender from the other side, from the vantage point of the "other America," is crucial, both because it challenges conventional and universalist assumptions about gender developed in Anglo-American scholarship and because the "other America" has been badly served by reductionistic views of how gender operates there.

After reading *Gender's Place*, you can say goodbye to "Marianismo" forever, with its image of the passive and subjugated Latin American woman whose agency is limited to the veneration of the Virgin Mary. You can also say goodbye, once and for all, to a host of other facile stereotypes about the place of gender across the border. In this volume, gender is never bland. *Siempre pica un poco.* Be careful. Gender's got spice here. And the reason is that gender matters in Latin America. It matters deeply, and the diverse group of observers of everyday gender relations included in this volume, who take us on a breathless journey from Venezuela to Ecuador to Nicaragua to Chile to Mexico to Peru to Bolivia to Brazil to the open wound of the U.S.–Mexico border, show us why and how in ways that are fresh and compelling. This relocation of gender happens in tandem with the placing of previously unseen and unheard protagonists on the stage of ethnography, protagonists who include rural Nicaraguan women, Andean women leaders, and Mexican "women of the night," as well as men struggling with the inherited notions of manhood, whether in the Ecuadorian Andes or the urban spaces occupied by Brazilian *travestis*. Ultimately, the various contributors to *Gender's Place* are working to *desalambrar*, to remove the fences around our preconceived notions of gender in Latin America, in hopes of revealing fresh possibilities for gendered interpretations of social reality that will reinvigorate feminist ethnography.

Given the already wide scope of the book and the vast territory covered, to have also included a chapter on Cuba would have been demanding too much. But in Latin America there is no country that has made as much an impact on the rethinking of the complexities of gender as has Cuba since the revolution of 1959. Seeking to do away with the image of the island as a backyard, emasculated colony of the United States, Cuba sought to carry out a far-reaching feminist revolution within the revolution, legislating gender equality, educating women in the professions, teaching men about the need for them to share domestic responsibilities, and bringing women into the work force in large numbers through the institutionalization of child care. While many positive changes took place in Cuba, it was years before the ideal of the macho revolutionary hero could be deconstructed without fear, leading to the fierce repression of gay men in the early years of the revolution, and the rise of a double-day pattern of labor for women, working now full-time both in the home and in the public sphere.

Since the fall of communism in Russia and Eastern Europe, gender relations have grown yet more contradictory, as Cuba searches for ways to
maintain its revolution through the expansion of tourism, which in turn
has led to the growth of both the female and male sex trade on the island.
It is a painful moment in Cuba's revolutionary history, as utopian dreams
fade and new dreams have yet to emerge. And this has profound repercus-
sions for all of Latin America. For the "other America," Cuba offered
a model of liberation in its brave and bold effort to build a socialist para-
dise ninety miles away from the North American superpower, which finds
the island's challenge to its hegemony so threatening that it continues to
maintain an economic embargo against it.

But perhaps it is a good thing that Cuba isn't included as a subject of
study in this book. Such an exclusion serves to remind those of us whose
encounters with the island have turned us into Cuba addicts or *cubanólo-
gos*, who can't help making the case for Cuban exceptionalism, that revolutionary discourse and practice have sprouted many seeds in Latin America,
from liberation theology to pedagogies of the oppressed to new song movements. After all, the very concept of *desalambrar*, of radical unfencing, that
anchors this book comes from a Uruguayan singer.

The three editors of this book, Janise Hurtig, Rosario Montoya, and Lessie
Jo Frazier were graduate students of anthropology at the University of
Michigan and so I knew them all, more or less well, in their formative intellectual years as budding feminist ethnographers. I am moved that they hon-
ored my presence in their lives, as the full-fledged scholars they are now, by
asking me to write this preface.

They also knew me in the years when I was formulating my own position as a feminist ethnographer while researching and writing *Translated
Woman: Crossing the Border with Esperanza's Story* (1993) and *Women
Writing Culture* (1995), an anthology of feminist ethnographic writing,
which I co-edited with Deborah Gordon, that began as a graduate course
and conference held at the University of Michigan. I went to graduate
school in the late 1970s and early 1980s, a period when feminist anthropology was a nascent subfield of our discipline and programs and departments in women's studies didn't yet exist. As a beginning professor,
I needed to scramble to teach myself about approaches to women and gender while simultaneously teaching my students. In contrast, the three editors of this book received their graduate training in the late 1980s and early
1990s, when feminist anthropology was in its heyday and supported by the
infrastructure of strong women's studies programs around the country, with

the University of Michigan leading the way in its commitment to women's studies.

Although we coincided at the University of Michigan and in various ways share a "Michigan approach" to the study of gender, the generational difference between us is dramatic and I think it accounts for the confidence and ease with which the editors approach the study of gender and pursue their feminist anthropologies. It is satisfying to see them not needing to justify intellectually the importance of gender, not needing to remind their readers that feminist anthropology offers an emancipatory agenda that includes both women and men, not needing to worry about being called strident because they draw attention to the connections between gender, race, class, and state power.

The editors are, indeed, the beneficiaries of the feminist movement, which gave gender a place and a home in the academy, a place and a home increasingly under threat. The threat comes not only from feminist backlashes of various sorts, within and beyond the academy, but from the very success that feminist mainstreaming has had in the world at large, which in some cases has led to the misuse and misappropriation of the concept of gender for the support of old oppressions packaged in new ways.

It is of key importance that scholars continue to work on articulating emancipatory agendas when they focus on gender. This book offers several such agendas and gives hope that scholarship, and feminist anthropologies in particular, can play an active part in positively transforming the everyday social construction of reality by helping us to see it in ways we had never seen it before.

Introduction ❧

A Desalambrar: Unfencing Gender's Place in Research on Latin America

Janise Hurtig, Rosario Montoya, Lessie Jo Frazier

This volume brings together recent work by feminist anthropologists of Latin America whose research addresses the interplay of gender, place, and power. As anthropologists, we find "place" to be a useful conceptual vehicle for emphasizing the cultural, social, and historical specificity of gender relations and ideologies. As feminists, we recognize the necessity of incorporating gender—understood as an axis of social power (McDowell 1999) and a category of analysis—to examine how specific places operate as sites of cultural production, institutional inequalities, or grassroots struggles. In other words, the chapters in this volume theorize specific social practices, cultural constructions, identify formations, and historical transformations *through* the inter-relationship of gender and place.

Working out of different strands within feminist anthropology, the chapters in this book examine how power, authority, and meaning are gendered, structured, produced, and contested in particular places: within the institutional realms of the state (Frazier, Hurtig, de la Cadena, Alonso, Paulson) and hacienda (Lyons); in the spaces created by political struggle or economic change (Cervone, Lamas, Klein, Ortiz, Montoya); as well as in the informal arenas of house and street (Montoya, Hurtig, Klein, Lamas).

Of course, feminist scholarship is not simply scholarship on the topic of gender, broadly writ; it is at the same time *critical* scholarship, scholarship that asserts a reflexive[1] stance toward one's social position as researcher, the frameworks one employs, and the images and discourses one deploys. Thus the intention of this volume is twofold: to contribute to a growing body of research on gender, power, and change in Latin America while also questioning and proposing alternatives to many of the guiding theories, formative concepts, and "master narratives" that have informed existing research.

The development of feminist research on Latin America has been impressive. What began as a narrow, two-pronged field, in which most research fell either into the economic realm of women and (under)development studies or the cultural realm of gender relations and ideologies/representations, has expanded to include studies of women's political involvement, their roles in grassroots struggles and so-called new social movements, as well as research on state formation and the effects of gender imagery in reproducing or transforming structures of power. The proliferation of personal narrative and testimonial-based studies and participatory action research projects have contributed to the field's inclusion of Latin American women as research partners and co-authors of their own experience, and not just as objects of academic research (see Behar 1993). Nonetheless, the field still suffers from a theoretical/conceptual split between materialist and culturalist approaches, signaled by June Nash in the 1980s (1986, 14). Nash urged researchers to move beyond the limitations of this split by integrating consciousness, culture, and material conditions in their examination of gender (1986; see also Melhuus and Stølen 1996). This volume invokes Nash's call for a new perspective. By drawing on the simultaneously spatial, social, and symbolic implications of "place" (Pratt 1998; Massey 1994), we aim to invigorate gender research in Latin America by *desalambrando,* or tearing down, the very distinction between the material and the cultural—demonstrating ethnographically the materiality of gender ideologies and the cultural force of gendered socioeconomic relations.

WHY "DESALAMBRAR"?

In developing a guiding premise for this volume we have found the concept of *desalambrar* (to undo or overrun fences) particularly suggestive. Desalambrar was introduced into the language of the Latin American New Left of the 1960s and 1970s by Daniel Viglietti, a Uruguayan singer and composer associated with the New Song Movement in Latin America. His song, "*A Desalambrar,*" urged Latin Americans to "tear down" the fences that kept them divided from one another and that cordoned off elite and foreign property in Latin America from the common people who worked on and produced wealth from those properties:

Yo pregunto a los presentes   I ask those present
si no se han peosto a pensar   if you not have stopped to think

que esta tierra es de nosotros	that the land belongs to us
y no del que tonga más.	and not to those who have more.
Yo pregunto si en la tierra	I ask if you have never thought
nunca habrá pensado usted	about the land
que si las manos son nuestras	that if the hands are ours
es neustro lo que nos dé.	what they give us is ours.
A desalambrar, a desalambrar,	Tear down the fences, Tear down the fences,
que la tierra es neustra,	Because the land is ours,
es tuya y de aquél,	it is yours and theirs,
de Pedro y Maria,	It is Pedro's and Maria's
de Juan y José.	and Juan's and José's.

Like dependency theory, liberation theology, pedagogies of liberation, and *testimonio* (testimonial literature), desalambrar is a regional model that emerged from the Latin American tradition of combining political–economic analysis with radical critique and practice. Desalambrar implies a praxis that both analyzes national configurations of Latin America's historical dependency and instigates emancipatory action directed toward class and national liberation. As such, it was an integral part of and consistent with the social movements that shook Latin America in the 1960s and 1970s.

In one sense, this book recognizes and actively engages with the important contribution that Latin American emancipatory projects, figured by the notion of desalambrar, have made to feminist theory and practice. For instance, current popular education and participatory action research projects addressing grassroots women's issues employ liberationist pedagogies inspired by radical educators such as Paulo Freire and popular theater workers such as Augusto Boal. Another example of this influence are the myriad feminist and women's movements across Latin America that developed out of revolutionary or Left party movements, albeit at times in reaction to the sexism of those movements (Stephen 1997; Radcliffe and Westwood 1993).

At the same time we see this volume as expanding the concept of desalambrar by examining how gender may be mobilized in, or impinge upon, the restructuring of discourses, practices, institutions, and communities. In addition we are turning the notion of desalambrar on itself, appropriating the term to argue that gender analysis can take apart and transform old analytic terrains—including those terrains upon which the notion of desalambrar itself was constructed. The gendering of desalambrar disrupts the unity of the androcentric subject assumed by most prevailing models of social change. It also complicates those models by forcing the recognition of multiple, gendered subjects whose activities and knowledge operate on history from the margins. Moreover, by retaining the notion of desalambrar—with its close attention to the role of structural factors in creating and sustaining relations of power—we avoid reducing subjectivity to (individual) identity. Instead, we examine the conditions of possibility for subjectivity by positing structures, and not identity per se, as both the site and object of struggle.

Once desalambrar is conceptually liberated from its purely class moorings, it can be used to dismantle other reified academic categories such as institutions and ideologies by exposing the multiple practices that strategically and continuously make, remake, and undo these terrains. Nonetheless, while desalambrar allows us to see the emancipatory potential of ordinary people's gendered practices, we must also be keenly aware of the hegemonic force of gender ideologies in legitimizing and naturalizing structures of power, which is to say their role in *alambrando*—fencing in, constraining and restraining—possibilities of change. The chapters in this volume exemplify how gendered ethnographic accounts can produce critical analyses toward social change that effectively temper, *without muting*, the Left's historical utopian impulses[2]—a political stance that is particularly urgent given this profoundly dystopian moment in Latin America and the world at large. In this volume, we work toward this goal by elucidating the gendered dynamics of localities as specific sites of power and by pointing out both possibilities and constraints to transformation that emerge thereof (see Dirlik 1996, 22).

WHY ANTHROPOLOGY?

It may seem contradictory to the transgressive, contestatory spirit of desalambrar to have deliberately compiled a volume of feminist research in Latin America that works out of a single discipline, that of Anthropology. After all, in the United States and in Latin America, feminist scholarship— and its institutional embodiment in women's studies programs and centers of gender research—has been insistently interdisciplinary, at times even antidisciplinary. Indeed, one of the contributions of feminism to critical research has been to challenge disciplinary boundaries, boundaries that "discipline" researchers to think narrowly and write exclusively. In many ways, the chapters in this volume reflect the turn toward interdisciplinary research that feminism has provoked. For instance, each chapter examines the construction and contestation of gender relations and ideologies *historically*—as emergent in social practices and not fixed in eternal structures or cultural patterns. While the chapters situate gender historically, they also draw on theories of gender and place developed by feminist and other critical scholars occupying diverse disciplinary locations, including literary criticism, philosophy, geography, and sociology. Two of the volume's final commentaries further this engagement by interrogating the possibilities and limits of interdisciplinary dialogues.

While feminist anthropology becomes increasingly interdisciplinary and continues to challenge many of the theoretical foundations, epistemological suppositions, and representational conventions of mainstream anthropology, it also builds upon certain disciplinary tendencies we consider to be the basis for feminist anthropology's contribution to gender research in Latin America. One of the tendencies within anthropology informing the chapters in this volume is its emphasis on the historical and cultural specificity of *local* practices and meanings. At its critical best, feminist anthropological research in Latin America draws on this disciplinary orientation to disrupt the presumed unity of gender relations and ideologies across the continent. The propensity of much current research to succumb to the "charisma" of globalism, as rendered through the opposition of "global forces" versus "local places" (Tsing 2000), similarly benefits from an anthropological insistence that "the cultural processes of all 'place' making and all 'force' making are *both* local and global, that is, both socially and culturally particular and productive of widely spreading interactions" (352).

Anthropology's focus on social organization and cultural production within specific localities cannot be separated from its grounding in participant-observation-based ethnography—a methodological orientation based on daily, intimate, long-term engagement with a community that is eminently suitable to describing social practices and cultural forms such as gender, and asking *how* they are reproduced and transformed (Foley, Levinson and Hurtig 2002). The chapters in this volume are all based on the ethnographer's commitment to building long-term social relations in particular locales as a methodological principle. In most cases, this commitment is coupled with a second methodological and political principle: sharing vulnerability with our ethnographic "subjects" by revealing our presence in the research process and in the questions we choose to address (cf. Behar 1993, 1996; Harding 1987). Even historical ethnography, in its approach to the archive as a field site, reflects this particular sensibility for engaging with processes in the past (see Alonso, Frazier, and de la Cadena in this volume) and foregrounds a dialectical tension between archive and field site.

A related quality of most anthropological inquiry is its insistence on the legitimacy of all human experiences and ways of knowing—a perspective that anthropology has managed to sustain, albeit often in tension with, the enduring Enlightenment tendency to explain (away) cultural differences reductively through the deployment of master theories of universal human processes. Though mainstream anthropology has often been "malestream" anthropology—deriving claims about whole societies from the points of view of its adult male members (Reiter 1975; Di Leonardo 1991)—the anthropological appreciation of multiple realities and epistemologies lends itself to feminist inquiry concerned with the variability and fluidity of gendered subjectivities.

Yet another, equally definitive, dimension of anthropological inquiry that serves to put the local in perspective is its theoretical and methodological "thickness"[3]—the production of understanding "through richness, texture, and detail, rather than parsimony, refinement, and... elegance" (Ortner 1995, 174). Without ascribing to the aesthetics of past anthropological holism that proposed to use ethnographic thickness to describe cultures as integrated systems, we do seek to rescue from that holistic vision an approach to social research based on the premise that all realms of social and cultural life are intimately interconnected and intercontextualized, and should be studied as such. The notion of "place"—referring to networks of meaningful (if generally unequal) social relations—is an effective conceptual vehicle for linking cultural specificity and ethnographic thickness. Moreover, ethnographic thickness is particularly relevant to feminist scholarship of Latin America, bridging the thematic and theoretical gaps between regional literatures we described earlier (see Nash 1986) by insisting that we theorize gender as simultaneously material and cultural, social and symbolic.

WHY GENDER?

The impetus for the volume emerged in the aftermath of an academic panel that we, the three editors, organized for a Latin American Studies Association meeting. The title of the panel was "Gender's Place: *A desalambrar y reestructurar instituciones e ideologías a fines del siglo XX.*" We were all recently returned from lengthy fieldwork stints in various places across Latin America: northern Chile, southwestern Nicaragua, Andean Venezuela; and we were caught in that torturous ethnographic moment when one clings to the ground, savoring the rich mundane stories of daily life, repeating the gossip over and over again, trying to grasp the larger picture while resisting the generalizations, the concepts, the models that seemed to pull us away from the conditions of people's lives even as they tempted us with explanations.

Despite our very different research projects, the common theme that asserted itself in all of our stories was the prevalence, persistence, and power of gender. The panel commentator, Ileana Rodríguez, described our papers as trying "to unravel how gender governs the social and symbolic order." Indeed, gender seemed to be a central organizing principle animating and cutting across all aspects of cultural life and social inequalities. As Melhuus and Stølen note in their introduction to a volume on gender imagery in Latin America, "gender differences in Latin America appear to take on a commanding significance in the conceptualization of difference and *the ordering of inequalities other than gender*" (1996, 2; our emphasis). This point bears emphasizing: Regardless of the explicit subject matter of our research, our shared concern with the reproduction of social inequalities and the possibilities of their contestation led us inevitably and necessarily to questions of gender. We thus came to recognize gender's place in research on Latin America as extending into but also far beyond the conventional thematic terrains of women's subordination; the social construction of femininity and masculinity; or sex(uality), marriage, and reproduction. At the same time, by locating gender at the center of our analyses, we found that these topics—traditionally ghettoized in social research by their very constitution as "gender/women's studies"—became central to understanding broader social questions, as the chapters in this book illustrate.

Our initial efforts to document and decode the subtle forces of gender were informed by an insistence on theorizing from the ground up and fueled by the lingering immediacy of our field experiences. This immediacy in turn sustained our sense of the epistemological and theoretical legitimacy of local forms of knowledge and a skepticism toward the presumed institutional legitimacy of North American and European analytic categories and concepts—a perspective expressed symbolically in our bilingual panel title that incorporated an indigenous concept (desalambrar) into our theoretical frame. In addition, our sense of what constituted an adequate understanding of the social and symbolic power of gender was informed by a shared feminist epistemology that proposes to assess analytic and interpretive endeavors in terms of their usefulness to meaningful social change (Mies 1983). It was this sense of adequacy that pushed us beyond a phenomenological recognition of the predominance of gender as an organizing principle of social life, to a critical examination of the efficacy of gender as a vehicle for constructing and representing myriad terms of social inequality. Our efforts to theorize gender inequalities, without sacrificing their specificity to the universalizing impulse of so many master theories, led us to theorize gender in relation to place.

WHY "PLACE"?

If we were going to usefully theorize the workings of gender as a technology of power (de Lauretis 1987), then we had to get past our enchantment with the ubiquity of gender across Latin America and attend to the specificity of gender arrangements, ideologies, and imageries in particular places. As we moved from telling stories, to trying to interpret them, to locating them within broader historical and cultural frames, we encountered two limitations within the literature. First, most existing accounts that addressed gender in Latin America did not seem to fully illuminate questions of gender in the places we were working. This is not to say those studies were wrong, but they were about different people in different places engaged in distinct struggles. Second, we found that explanations of women's oppression in Latin America often resorted to, and thus assumed the universality of, Euro-U.S. cultural and social binarisms: They situated gender in terms of organizational oppositions (such as "public/domestic") or cultural constructs (such as "machismo/marianismo") whose significance was assumed to be constant across history and geography. Our commitment to theorizing gender in relation to place—both how places create and constrain historically specific gender inequalities and how gendered social processes create specific kinds of places—provided the critical momentum to desalambrar gender research from the constraints of regional generalizations and categorical imperialisms.

This volume exemplifies the critical usefulness of engaging gender and place. First, the reader will find in these chapters "thick" ethnographic accounts of the specificity of gender arrangements and ideologies in particular places. The intention of this kind of work is not exhaustiveness or dense contextualization (Ortner 1995, 174) per se; but rather to produce the kind of situated analysis that can offer "operationalizable" (Strathern 1987) insights accessible to scholars, community activists, and policymakers alike. The emphasis of this volume on ethnographic case studies in which the meanings and import of gender are shown to be locally specific challenges the tendency to assume certain commonalities to gendered social roles, relations, ideologies, and identities across the continent of Latin America.[4] Each of the authors in this volume goes beyond feminist critique to offer alternative models of local forms of gender. Many also show how those local notions are deployed in the legitimation and reproduction of other social inequalities.

Another way in which the interplay of gender and place has a critical effect upon social theory is that it calls into question notions of context and contextualization—a mystified analytic enterprise that belies the ways "our understandings of social relations and history, understandings which constitute the fabric of such context ... [are] themselves fragile intellectual constructs posing as robust realities obvious to our contextualizing gaze" (Taussig 1992, 45). Thus it is "the fabric of context" that most cries out for analytic interpretation (Taussig 1992, 45; see also Egmond and Mason 1997). As the chapters in this book demonstrate, by examining the dialectics of gender and place the fixity of each is effectively dissolved and their conditions historicized.

In this volume we are mindful as well of the gendered nature of "context." In conventional social science writing, women and the domestic often function as supporting cast and scenery respectively, while men play the role of historical actors. The chapters in this volume write against such a subtext by foregrounding the agency of women, social marginals, and oppressed communities in struggling against and for particular constructions of place. By noting how gender dynamics are intrinsic to the construction of places, we are also forced to recognize the contingency of places and the way in which they are constructed by webs of ever-shifting social relations of power—replacing the certainty of context with the instability of place. This opens up our examination of place to the workings—and potential dismantling—of hegemonic forces (cf. Massey 1994).

At the time we began conceptualizing this book, the hegemony of the global—as a model for political–economic restructuring and a spatial metaphor for conceptualizing those changes—had not yet gained the dominant place in academic discourse it currently occupies. Given the conceptual emphasis throughout this volume on the enduring distinctiveness and force of the local, it seems incumbent upon us to address the place of the local in relation to the global. Where it is relevant to issues being addressed, individual authors attend to regional, national, and transnational processes and cultural forms as they operate in particular locales. In each chapter, however, the authors demonstrate that the impact and shape of translocal processes varies depending on local cultural forms, and often on the personal, familial, class, and gender locations of the subjects. The intention here is not merely to use the local as a lens onto the national or transnational, but to demonstrate ethnographically how society, the state, and global processes are always located—that is, constituted by and through daily practices and experiences in particular places (cf. Tsing 2000). In this sense, our effort is to grasp the dynamics of social relations not easily contained in the dichotomous geographical imaginaries of "local" versus "global." It is this kind of critical approach to the specific, located incarnations of national and global forces that both illustrates and makes possible the intervention of ordinary people in processes of political–economic reconfiguration and social change.

CONTRIBUTORS

In compiling this volume, we have made an effort to transgress established frameworks—academic practices, intellectual flows of discussion and debate—that reproduce intellectual hierarchies and exclusionary geographies of knowledge within the academy. One way we do this is by bringing together the works of scholars who are rarely read as part of a common dialogue. The chapters in the volume integrate the research concerns of scholars of Latin America based in the United States with those based in Latin America whose work may be relatively unknown in this country. The book also brings together in dialogue established scholars with those just beginning their scholarly endeavors. Contributors work in scholarly settings ranging from universities to research institutes to non-governmental organizations. Recognizing that institutional settings create cultures of research and understanding that both promote and limit the nature and terms of social inquiry, we have compiled this disparate volume in an effort to tear down the institutional fences of knowledge production that constrain the possibilities for new ways of thinking critically about gender's place in Latin America.

Part 1: Gendered Knowledge in Particular Places

Part 1 explores how gendered hierarchies and meanings, and the possibilities for their transformation, are constituted through gendered ways of knowing in common places: a Venezuelan secondary school classroom, highland Ecuadorian haciendas, and the spaces of house and street in a Nicaraguan village. Recognition of the situatedness of signifying, learning, and knowing in time and space compels these authors to build their analytic concepts and/or models out of local categories. While these chapters offer strong cases for the potential of situated research to desalambrar terrains of research obscured by Euro-U.S. categories, they also point to the challenges of such an endeavor, given that local forms of knowledge may be—wittingly or not—complicit with the very structures of power that inscribe gender as an inequality.

In "Debating Woman," Janise Hurtig focuses on Venezuelan students' preparation for and enactment of a debate on "the role of Venezuelan women in society," to consider the relationship between gendered forms of learning and learning about women/gender. The chapter explores how students, engaging in spatial practices called "learning," collectively defined their social selves while structuring the conditions of their lives. Hurtig's essay complicates feminist approaches to schooling by showing how educational activities are gendered through and through, and how students' uses of their implicit knowledge about their gendered worlds can be obscured by conventional educational research frameworks.

Barry Lyons's chapter, "To Act Like a Man," explores the varied meanings of maleness in Ecuadoran Quichua speakers' folk tales and accounts of life on highland haciendas. Lyons's close reading of contexts and notions of resistance among Quichua speakers defamiliarizes this concept, the indiscriminate and widespread use of which in anthropology has become all too problematic (Abu-Lughod 1990; Ortner 1995). By working from the ground up, Lyons shows that resistance in the Ecuadorian Andes is centrally informed by gender ideologies, and that its complex meanings cannot be understood outside a particular historical and cultural context. He prompts anthropologists to be self-conscious about their uses of this concept when he suggests that our own notions of resistance can be subjected to the same sort of analysis (46).

In "Women's Sexuality, Knowledge, and Agency in Rural Nicaragua," Rosario Montoya argues that women's knowledge about the workings of local gender ideologies allowed them to craft the conditions of possibility for greater rights to sexual and conjugal choice across their life cycle. Montoya argues for an examination of women's practices from their own points of view as subordinate actors knowledgeable about their conditions of existence. In so doing, Montoya's analysis furthers the project of *desalambrar* by shedding light on the force and significance of "domestic" practices that would be obscured by the imposition of categories such as public or private upon women's experiences and by questioning conventions of much feminist research that associate Latin American women's historical agency almost exclusively with their incursion into the public domain.

Part 2: Gender's Place in Reproducing and Challenging Institutions and Ideologies

Part 2 develops the usefulness of gender as a category and feminism as a kind of critical social theory to understand the workings of institutions and ideologies that have shaped political dynamics in the region. The essays in this section deal with various aspects of the state, including social policy and development projects, public health, legal frameworks for social cohabitation, and projects for remaking the nation-state. These ethnographic case studies suggest that, while state projects and institutions set the parameters within which actors move and struggle, these are, in turn, always shaped by the intersection of dominant conceptions of gender relations, race/ethnicity, and class. The analytic distinction between "ideologies" and "institutions" of the state is thus dissolved in (and through) practice. These studies also show what recent ethnographies of the state have argued (Tsing 1994; Joseph and Nugent 1994; Taussig 1997): that the state is not only present in its formal institutions, but is also incarnated in locally occurring negotiations, as well as in spaces that appear remote or unconnected to state apparati.

Lessie Jo Frazier's chapter, "Forging Democracy and Locality," analyzes the efforts of Chilean human rights activists to renegotiate their relation to the newly democratic state through a pilot mental health and human rights reparation program. By examining the use of local knowledge to dismantle structures of power, Frazier sheds light on the gendered processes though which local actors attempt to negotiate the possibility of community and hegemonic boundaries of place. Frazier shows that in this process, they create "paradoxical places" (Rose 1993; Desbiens 1999) of which the complex array of relations and meanings exceeds the bounds of singular frameworks, demanding a more multidimensional analysis of power and place.

As Frazier's chapter suggests, the life of the nation-state is constituted in particular and quotidian locales through the agency and ideas of different social sectors attempting to negotiate and participate in shaping hegemonic projects. These negotiations turn, in part, around state policies regarding violence in society. With Ana Alonso's chapter, "What the Strong Owe to the Weak," we move from the confrontation of the legacies of state violence to the regulation of domestic violence. Alonso focuses on the historical moment of the emergence of a modern liberal state in Mexico to show how legislation criminalizing domestic violence that would appear to work in favor of women actually contributed to their ongoing oppression. Thus, women who used the legal system to challenge their husbands' abusive power also affirmed hegemonic constructions of femininity and, hence, their submission. Alonso's essay identifies not only how gender inequalities were sustained through "rational" policies of state formation, but also how women (and men) contested the use of violence to uphold patriarchal power in the domestic sphere. In so doing, she writes against the narrative of modernization that underpins Anglo images of Mexicans and women of color by making "the North the site of reason and the South the site of irrational male violence [and] female passivity" (Alonso, this volume, 132). At the same time, Alonso challenges the Foucauldian opposition between the productive and prohibitory character of power by showing that legal prohibitions were indeed productive of subjectivities, normative classifications, and forms of power/knowledge.

Susan Paulson's chapter, "Placing Gender and Ethnicity," looks at the complex relations between various Bolivian scholars and state policymakers as they sought to incorporate gender and ethnicity as analytic categories for government-funded research and policy initiatives in the emerging neoliberal political economy. In the context of these negotiations around the institutionalization of analytic categories, a new possibility emerged: the concept of *transversalización,* a pluriethnic and gender sensitive vision for citizenship and political agency that offers an analytic tool to eventually desalambrar the structures of knowledge and power. In this short-term case, the politics surrounding the implementation of this concept, including the embeddedness of Left intellectuals themselves in elite positions, kept these academic and governmental actors from living up to the potential of their own insights.

In Marisol de la Cadena's "The Political Tensions of Representation and Misrepresentations," the focus of analysis shifts from organizations (Frazier) and state institutions (Alonso) to discourses without a direct connection with, but very clearly central to, state projects. De la Cadena looks at gendered constructions of ethnicity deployed by the Cuzco elite during two periods of nation formation and the socioeconomic policies that were developed in terms of and legitimized by those constructions. Through a focus on market women's redefinition of elite ethnic categories, de la Cadena challenges analytic tendencies to draw cultural boundaries between present day "Indians" and "*mestizos.*" At the same time, her case study reminds us that subaltern challenges never occur outside webs of power: as market women defined themselves as *mujeres de respeto,* they upheld rather than challenged social hierarchies between "superiors" and "inferiors." Like Frazier, Alonso, and Paulson, de la Cadena underscores both the potential and the limitations of tearing apart social hierarchies defined —and countered—through the terms of dominant ideologies of race/ethnicity, sexuality/gender, and the political and social body.

Part 3: Gender in Movement(s)

Part 3 builds on themes of transformation presented in earlier chapters. The four case studies look at gender "in movement," addressing problematics of the relationship between gender and place in processes of geographic movement—such as globalization and migration—or social movement, such as grassroots struggles. These articles foreground the transformative force—and limits to transformation—posed by the agency of "marginal natives" vis-à-vis established structures of power. Whether they be indigenous women leaders in Ecuador, Latino women professionals at the U.S.–Mexico border, *travesti* sex workers in Brazil, or female sex workers in Mexico City, they all are engaged in social, economic, and/or political activities that require them to work within the constraints of the very gendered structures and meanings they are consciously seeking to tear down.

Three of the essays presented in this section (Cervone, Klein, and Lamas) resulted from collaborative efforts of the authors as feminist researchers with the communities whose struggles they document, each of the authors having taken part in participatory research projects aimed at assisting the organizational efforts and enhancing the institutional presence of these social groups. Emma Cervone's "Engendering Leadership" addresses the pathways through which women are entering the male space of political leadership among Ecuadorian Quichua Indians. Through an account of the barriers two women have had to surpass to achieve leadership positions Cervone describes the limited gender egalitarianism of a movement striving for ethnic revindication. Her comparative discussion of the two cases relates the different limitations and possibilities of their capacity to politicize gender identities to their distinct positions as ethnic leaders differently situated vis-à-vis *mestizo* society. Like Alonso and de la Cadena, Cervone shows that the struggle to create spaces for leadership in political terrains saturated with discourses of modernization is in part a struggle to redefine the relationship between race, ethnicity, and gender in the political arena.

Victor Ortiz's "Latinas on the Border" echoes key themes in Cervone's chapter, such as the entry of women into male spheres and the differences that women's social and political positions make to their roles as leaders. Ortiz compares two Latina professional women's strategic responses to "the opportunities and challenges provided by globalization" in the specific border site of El Paso. By looking at women as social actors operating in a sphere usually ascribed to men, Ortiz's chapter challenges conventional views of the relationship between capitalism, gender, class, and ethnicity that totalize "women" as a class. It also challenges increasingly dominant notions of the relationship between the local and the global as places ascribed with differential degrees of impact. This gendered relationship—in which the encompassing global form is attributed with more determinative weight than the local—is shown by Ortiz to be in fact very complex and intertwined.

The section's subsequent chapters extend the discussion of ideologies of the street introduced earlier by Montoya and discussions of the space of the city brought forth by Ortiz, Frazier, and de la Cadena. In "Making a Scene" Klein examines the efforts of transgendered sex professionals to transform existing terms of political action, construct a new political space, and thereby facilitate their own political action, in the process (re)constituting themselves as political actors. Klein's case study echoes many of this volume's chapters (Lyons, Montoya, Alonso, de la Cadena) by showing how gender-transgressive practices can become political challenges, but also how existing political spaces impose certain forms on those challenges.

In "By Night, a Street Rite," Marta Lamas presents the efforts of women sex workers in Mexico City to articulate their understandings of their labor and to reclaim their place of work, *el punto* (the point, or corner), through rituals. Because she works out of a participatory, activist ethnography, her method and analysis insistently incorporate the knowledge of the women she is "studying." While pointing to the cultural criticism implicit in rituals of the street, Lamas concludes that this work—the sex workers' critique and the anthropologist's analysis—is not sufficient, but requires a transformative political project to ultimately dismantle old terrains of power; she offers an interpretation that addresses issues relevant to the women she is working with and Mexican society in general. We see this as Lamas's attempt to offer interpretations that are "operationalizable"—useful as tools for political intervention (Strathern 1987). In so doing, she tears down methodological and epistemological fences determining what constitutes knowledge and legitimate research.

Part 4: Critical Commentaries

Part 4 offers three essays that elaborate on the concept of desalambrar as a useful tool, not only for thinking about configurations and contestations of power in "the field," but also in "the academy." The section's first essay, "Against Marianismo," is an interdisciplinary essay by historian Marysa Navarro that reflects on the politics of analytic categories across the Americas and across disciplines. Navarro's commentary traces the contemporary intellectual history of "marianismo," a concept that—whether scholars agree with it or not—has framed much of feminist scholarship on Latin America (including U.S. Latinos) produced in the United States and Europe. Her essay carries on the work of postcolonial scholars who foreground the limits of academic sisterhood and feminist scholarship as interor transdisciplinary endeavors.

By challenging us to confront the geopolitics of theory and the simultaneously tense and productive dynamics between disciplines in feminist research on Latin America, Navarro's essay compliments the final two commentaries. In "Understanding Gender in Latin America," Sonia Montecino, an anthropologist and creative writer at the University of Chile, offers the term *torcer* (to twist or bend) as a complimentary concept to desalambrar. She uses *torcer* to think about how feminist scholars based in Latin America appropriate and creatively refigure analytic tools coming from a variety of places, but too often overvalorized when coming from "the North."

In the final commentary, "Local/Global: A View from Geography," Altha Cravey, a feminist geographer of the Mexico–U.S. border, reflects on issues of scale in the potentially fruitful dialogue between feminist geographers and anthropologists. By encouraging feminist ethnographers to "focus more attention on the ways that ordinary people shape worlds that extend beyond their everyday routines, even if these routines appear to be predominantly local," Cravey (this volume, 283–284) offers a method for feminist anthropologies of place to deflate "globalist fantasies" (Tsing 2000) by unfencing conventions of analytic scale.[5]

If anthropology in general has the potential to critically illuminate transformative social and cultural operations, then feminist anthropology in particular has that capacity. By taking an anthropological approach to gender's place, the chapters in this volume direct our critical vision toward those social actors whose perspectives and actions can potentially disrupt fields of power precisely because of their locations on the margins of geopolitical, social, and intellectual centers. Here we want to distinguish our position from the tendency to romanticize the margins as a site of resistance, thereby "[acquiescing] to the centrism of another position" (Youngbear-Tibbetts 1998, 40). Rather, much as Marx argued that the way to "undo" the alienation of class is, as Terry Eagleton puts it, "to go, not around class, but somehow all the way through it and out the other side" (Eagleton 1990, 23), we argue that the way to undo (desalambrar) the alienation of gender is to locate and then "go all the way through" gendered power relations as they exist at this historical moment. This volume offers desalambrar as a praxis of mapping and dismantling relations of power by engaging the transformative potential of gender's place.

NOTES

1. Feminist philosopher Lois McNay describes reflexivity as "the critical awareness that arises from self-conscious relation with the other" (1999, 5).
2. See Castañeda (1994) for a discussion and critique of the Latin American Left's utopian projects.
3. The enduring value and changing significance of ethnographic thickness, a concept originally elaborated by Clifford Geertz (1973), is discussed by Sherry Ortner (1995).
4. See, for example, Jo Fisher (1993), especially pages 3–6.
5. See Marcus (1998) for a discussion of similar concerns within anthropology.

REFERENCES

Abu-Lughod, Lila. 1990. "The Romance of Resistance: Tracing Transformations of Power in Bedouin Women." *American Ethnologist* 17(1): 41–55.
Behar, Ruth. 1993. *Translated Woman: Crossing the Border with Esperanza's Story.* Boston: Beacon Press.

————. 1996. *The Vulnerable Observer: Anthropology that Breaks your Heart.* Boston: Beacon Press.

Castañeda, Jorge G. 1994. *Utopia Unarmed: The Latin American Left after the Cold War.* New York: Vintage Books.

de Lauretis, Teresa. 1987. *Technologies of Gender: Essays on Theory, Film, Fiction.* Bloomington: Indiana University Press.

Desbiens, Caroline. 1999. "Feminism 'in' Geography: Elsewhere, Beyond and the Politics of Paradoxical Space." *Gender, Place, and Culture* 6(2): 179–185.

Di Leonardo, Micaela. 1991. "Gender, Culture, and Political Economy: Feminist Anthropology in Historical Perspective." In *Gender at the Crossroads of Knowledge,* edited by Micaela Di Leonardo. Berkeley: University of California Press, 1–48.

Dirlik, Arif. 1996. "The Global in the Local." In *Global/Local: Cultural Production and the Transnational Imaginary,* edited by Rob Wilson and Wimal Dissanayake. Durham and London: Duke University Press, 21–45.

Eagleton, Terry. 1990. "Nationalism: Irony and Commitment." In *Nationalism, Colonialism, and Literature,* essays by Terry Eagleton, Fredric Jameson, and Edward Said. Minneapolis: University of Minnesota Press.

Egmond, Florike and Peter Mason. 1997. *The Mammoth and the Mouse: Microhistory and Morphology.* Baltimore: Johns Hopkins University Press.

Fisher, Jo. 1993. *Out of the Shadows: Women, Resistance and Politics in South America.* London: Latin America Bureau.

Foley, Doug, Bradley Levinson, and Janise Hurtig. 2002. "Anthropology Goes Inside: The New Educational Ethnography of Ethnicity and Gender." In *Review of Research in Education,* edited by Walter Secada. Washington, DC: AERA, 37–98.

Fuller, Norma. 2001. "The Anthropology of Gender in South America." *American Anthropological Association Newsletter,* May 16.

Geertz, Clifford. 1973. *The Interpretation of Cultures.* New York: Basic Books.

Harding, Sandra. 1987. "Introduction: Is There a Feminist Method?" In *Feminism and Methodology,* edited by Sandra Harding. Bloomington: Indiana University Press.

Joseph, Gilbert and Daniel Nugent. 1994. *Everyday Forms of State Formation: Revolution and the Negotiation of Rule in Modern Mexico.* Durham: Duke University Press.

Lamas, Marta, ed. 1996. *El género: La construcción cultural de la diferencia sexual.* Mexico D.F.: Programa Universitario de Estudios de Género, Universidad Nacional Autónoma de México.

Marcus, George E. 1998. *Ethnography in/of the World System: Ethnography through Thick and Thin.* Princeton: Princeton University Press.

Massey, Doreen. 1994. *Space, Place, and Gender.* Minneapolis: University of Minnesota Press.

McDowell, Linda. 1999. *Gender, Identity and Place: Understanding Feminist Geographies.* Minneapolis: University of Minnesota Press.

McNay, Lois. 2000. *Gender and Agency: Reconfiguring the Subject in Feminist and Social Theory.* Cambridge, UK: Polity Press.

Melhuus, Marit and Kristi Anne Stølen. 1996. *Machos, Mistresses, Madonnas: Contesting the Power of Latin American Gender Imagery.* London, Verso.

Mies, Maria. 1983. "Toward a Methodology for Feminist Research." In *Theories of Women's Studies,* edited by Gloria Bowles and Renate Duelli Klein. London: Routledge and Kegan Paul.

Montecino, Sonia and María Elena Boiser, eds. 1993. *Huellas: Seminario mujer y antropología.* Santiago: CEDEM.

Montecino, Sonia and Regina Rodríguez, eds. 1999. *Espejos y travesías: Antropología y mujer en los 90,* Ediciones de las Mujeres No. 16. Santiago: ISIS.

Montes, Soledad Gonzalez, ed. 1990. *Mujeres y relaciones de género en la antropología latinoamericana.* Mexico D.F.: El Colegio de Mexico.

Nash, June. 1986. "A Decade of Research on Gender in Latin America." In *Women and Social Change in Latin America,* edited by June Nash and Helen Safa. South Hadley: Westview Press.

Ortner, Sherry. 1995. "Resistance and the Problem of Ethnographic Refusal." *Comparative Studies in Society and History,* 37(1): 173–193.

Pratt, Geraldine. 1998. "Geographic Metaphors in Feminist Theory." In *Making Worlds: Gender, Metaphor, Materiality,* edited by Susan Hardy Aiken, Ann Brigham, Sallie A. Marston, and Penny Waterstone. Tucson: University of Arizona Press, 13–30.

Radcliffe, Sarah and Sallie Westwood. 1993. *"Viva": Women and Popular Protest in Latin America.* London: Routledge.

Reiter, Rayna. 1975. "Introduction." In *Toward an Anthropology of Women,* edited by Rayna Reiter. New York: Monthly Review Press, 11–19.

Rose, Gillian. 1993. *Feminism and Geography: The Limits of Geographical Knowledge.* Cambridge: Polity Press.

Stephen, Lynn. 1997. *Women and Social Movements in Latin America: Power from Below.* Austin: University of Texas Press.

Strathern, Marilyn. 1987. "An Awkward Relationship: The Case of Feminism and Anthropology." *Signs,* 12(3): 276–292.

Taussig, Michael. 1992. *The Nervous System.* New York: Routledge.

——. 1997. *The Magic of the State.* New York: Routledge.

Tsing, Anna. 1994. *In the Realm of the Diamond Queen.* Princeton: Princeton University Press.

——. 2000. "The Global Situation." *Cultural Anthropology* 15(3): 327–360.

1. Gendered Knowledge in Particular Places

1. Debating Women 〜

Gendered Lessons in a Venezuelan Classroom

Janise Hurtig

> *Spatial practices in fact secretly structure the determining conditions of social life.*
>
> —Michel de Certeau, *The Practice of Everyday Life*

INTRODUCTION

My field notes remind me that it was a Wednesday morning in March of 1992, six months into the school year and nine months into my ethnographic research in the Venezuelan town of Timotes.[1] I left the house early and joined the flow of students as they headed to the first-period class at Liceo Parra,[2] the municipal secondary school. Each student sporting the requisite national uniform—blue jeans and beige or white shirts depending on their grade, blue sweaters on chilly Andean days—they created a nearly monochrome stream of disciplined informality that briefly transformed the otherwise motley, sparsely inhabited, and principally adult space of the street. I was looking forward to attending the first class that day. Profesora Nilda Martínez,[3] the serious, soft-spoken young ninth-grade language and literature teacher, was holding a "*debate*" (debate in Spanish) on "the role of Venezuelan women in society," and she was expecting me to attend.

Timotes is an agricultural town and municipal capitol situated in a fertile valley in the Venezuelan Andes Since the 1950s, when the construction of the Transandean Highway connected Timotes to the country's economic and population centers to the east, the town's already lucrative horticultural industry had grown to serve the entire country. Compared to the string of small villages dotting the highway, Timotes stood out for being more afflu- ent and commercially vibrant, and thus perhaps less picturesque, than its neighbors. At the time of my stay in the early 1990s, the town's geographic remoteness and economic prosperity combined to buffer it from the eco- nomic, political, and social crises that had been spreading and deepening across most of the national landscape since the 1980s. As a prosperous agri- cultural town in one of Latin America's most urban countries, where decades of rapid modernization had been spurred by a tremendous petro- leum export economy now in crisis, Timotes was something of a national and regional anomaly (Hurtig 1998a,b).

For the students attending Liceo Parra, the paradox of their town's remote- ness and prosperity confused the legitimacy of the modern Venezuelan nation's rhetoric about the putative road to upward mobility in which an edu- cational trajectory from elementary through post-secondary education was mapped onto a spatial trajectory from the village to the city. It also made Timotense students' decisions about how to improve themselves—a crucial concern they were meant to take on in their self-construction as educated per- sons (Levinson, Foley, and Holland 1996)—less straightforward than might have been the case at other times. The local refrain about Timotes that "here nothing ever happens" was used by some students to convey their frustrated boredom with the provincial town and by others to express their appreciation for the tranquil if uneventful life it offered. The hopeful uncertainty of the future Timotes offered its youth also shook up local gendered conventions of coming of age that were meant to endow young men with ever-increasing mobility within and across local, regional, and even transnational spaces while inscribing young women in ever-constricting spheres of movement as they became mothers and wives (Hurtig 1998a; cf. Massey 1994, 179–180).

And Timotense youth knew that these conventions had been loosed from their moorings. Diana, a graduating secondary-school student, com- mented during an after-school discussion that "if you want to be someone in life you have to leave Timotes. You have to go to the city."[4] Meanwhile, Diana's classmate Ramón proclaimed that "here in Timotes we have every- thing." He planned to work in his brother's store, save his money, eventu- ally buy some land near his father's farm, build his house, and then have a family (Hurtig 1998a, 34). When Diana argued that Ramón was wasting his chances to make something of himself, Ramón's friend Miguel agreed that it was better to stay in Timotes: "Now, with the crisis, you never know if you'll be able to study in the university, let alone complete your pro- gram." Miguel's comment emboldened the usually taciturn Susana to speak up. "Well," she proclaimed, "I think one should go for a *carrera corta* [a short-track vocational degree] because, as my parents tell me, with the crisis you can't raise a family on one income anymore." A thick Andean silence followed Susana's remark, until Graciela murmured shyly, "That's what my mom says. But my dad won't let me go to Valera [the nearby industrial city] to study. He says it's too dangerous for a girl to leave home to study. He says girls who go to town to study acquire bad habits." Diana, her eyes fiery, snapped a response: "That's how Andean men are: very *machista*. They always want to keep women locked up, keep them from improving themselves. What's the point of going to school if afterwards we're just supposed to dedicate ourselves to the home?"

In this and other discussions I had with Timotense students over the course of the school year, I heard them render the process of becoming adults as scenarios that required integrating several life projects: completing an education, forming a family, embarking on a career. Imagining any of these scenarios entailed the projection into the future of a gendered social self. Not any self, but "a social 'me' defined here and now by the set of its sociological determinations" (Bisseret 1979, 33). But these determinations are not, I would argue, predetermined; rather, they are used, learned, created anew, and transformed in and through the social practices of daily life that occur in legitimizing spaces such as the classroom.

In this chapter I focus on the singular event of a ninth-grade classroom debate to consider how Timotense students, engaging in gendered spatial practices called "learning," defined their social selves while secretly structuring the conditions of their lives. As I trace some of the knotty, dangling, and unwoven threads of meaning the students spun out of this educational event, I want to remind the reader to distrust the notion that there is any clear path from "learning" to "knowing" to "doing." When I last visited Timotes, Diana was a mother of two and Susana was studying accounting through the local branch of the national open university, "in no hurry to become a mother, let alone get married," as she put it. Ramón was indeed working in his brother's store but, with the economic downturn, he wondered whether he would be able to buy a plot of land and contemplated a career in engineering. This chapter is concerned with how students like Diana, Susana, and Ramón implicitly deliberated over the possibilities of their lives as adult women and men as they explicitly debated "the role of Venezuelan women."

RE-PLACING LEARNING

Two weeks prior to the debate, Ms. Martínez had invited me to attend the event and help her evaluate the two debating groups' performance. When

eight students from one of the groups appeared at my doorstep the following week and asked if I would serve as their *asesora* (advisor) in preparing
for the debate, I went back to Ms. Martínez, got her permission to do so,
and then agreed with her that it would be inappropriate for me to also serve
as an evaluator. Instead, I would simply observe. But, she added as though
searching for a way to include me more actively in the event, I was welcome
to contribute to the discussion, ask questions, or correct any incorrect
information that the students provided.

I had visited Ms. Martínez's classes previously and had found her to be a
pleasant and thoughtful if rather conventional teacher. We had also struck
up conversations in town on several occasions, and each time she seemed to
lead the conversation toward the topic of women. I soon got the sense that
she saw me (the anthropologist who was commonly known to be studying
women or sexual relations) as a kindred spirit. In our discussions about
higher education and her own training to become a teacher, she generalized
from her personal experiences to the systematic discrimination of women
students by the university. On these matters she was always thoughtful and
circumspect. Thus, although she never told me so directly, it was my sense
that when Ms. Martínez organized the debate she fully intended to counter
the conspicuous absence of "women" in the secondary school curriculum.
That she chose to use the novel and potentially critical pedagogical form of
a debate to expose the students to a topic that, in and of itself, implied a
critique of the conventional curriculum was, I suspect, no coincidence.
Ms. Martínez, in her quiet and unobtrusive way, was something of a rebel.

This essay, however, is not an exploration of Ms. Martínez's political
motives or educational intentions, but rather an inquiry into the gendered
social conditions that the students drew upon and worked on as they prepared for and engaged in the debate. In framing my inquiry, I draw on
three key propositions laid out in the introduction to this volume. The first
proposition is that the conditions determining gender(ed) inequalities are,
to paraphrase the chapter's epigraph, secretly structured through spatial
practices. The second proposition is that feminist cultural analysis should
give an account of those practices in a theoretically oblique way that reveals
them as doing "critical ideological work" of gender (Poovey 1988, 2–3),
while contesting conventional cultural and social analyses that obscure
those practices as such. The feminist literary critic Mary Poovey uses the
term "ideological work" to refer both to "the work of ideology" and to "the
work of making ideology" that occurs through the "complex interaction
of images and social institutions" (Poovey 1988, 2). In thinking about
the debate as doing such ideological work of gender, I want to emphasize
the "two guises of ideology—its apparent coherence and authenticity, on

the one hand, and its internal instability and artificiality, on the other" (3). This instability is emergent in the communicative work people do as they negotiate their social relations and individual subjectivities in terms of those ideologies. Gender ideologies, that is, are always subject to debate. Furthermore, ideological formulations are uneven—incoherent and unstable—in part because they are differently experienced and understood by individuals depending on their social location (Poovey 1988). In other words, the very social inequalities that gender ideologies are meant to essentialize and thus fix, in fact contribute to their instability, their dynamism.

A feminist inquiry that approaches educational practices (such as the debate) as gendered ideological work thus entails tearing down myriad conceptual fences that have framed learning theory for centuries—fences built upon familiar Cartesian scaffolding that separates individual from society, mind from body, thought from action, and actors from their putative contexts. Each of these distinctions is supported by Enlightenment social theories that begin and end with the authentic individual as the unit of social analysis who is positioned within, but somehow separable from, an unproblematically observable context (cf. Taussig 1992, 45). By taking learning as an individual achievement and the horizons of that achievement as determined by the student's constitution and/or individual behavior (aptitudes, *habitus*, sex, ethnicity, class background, for instance), any exploration of learning as collective and contested gendered cultural activity is precluded. And by taking the learning "context" to be a static set of spatially, temporally, and socially delimited conditions separable from but determinative of the individual learner's constitution and/or behavior, any inquiry into historical process is foreclosed. As Ray McDermott puts it, "a static sense of context delivers a stable world" (1993, 282).

These conceptual mystifications, I fear, have guided much of the research on women/gender and education in Latin America, in which causal relationships between female students' school performance and their eventual location in the work force are invoked as explanations for the reproduction or transformation of gender inequalities, and in which educational texts and discourses are analyzed for their gender "content," as though gender ideologies are poured directly into the half-empty heads of students who then act accordingly or bristle and resist. Such research, that treats schooling as a process of mechanical reproduction and students as its passive objects, is limited in the social processes it can account for, and thus the paths toward change it can propose. The propitious tendency of recent anthropological studies of schooling in Latin America toward interpretive accounts of the cultural production of personhood through educational practices[5] is much more conducive to feminist research practice.

However, more is required to contest this positivist hegemony than the simple conceptual move of re-locating learning within the field of ethnographic examination; it is equally important that we situate ourselves as researchers within that field. The third proposition I draw upon from this volume's introduction is that we as researchers are always necessarily implicated in the ethnographic moment we study, such that attention to place, to "contextualization," as Michael Taussig proposes, "should be one that very consciously admits of our presence, our scrutinizing gaze, our social relationships..." (1992, 45). Such an admission, Taussig continues, "opens up to a science of mediations—neither Self nor Other but their mutual co-implicatedness." A statement that brings me to tears and laughter as I attempt to open the debate up to such a science and recall (with some embarrassment) that the debate's essentializing topic, "the role of Venezuelan women in society," was also at the heart of my own ethnographic inquiry. So that my thoroughly compromised participation in the debate as expert/teacher/observer (a product of which is this chapter) must be included in the expanding and ever-messier terrain of gendered ideological work that I am here calling "learning." If the reader finds in these pages little resolution of that mess, it is with the intention of encouraging you to subject the debate, as well as my interpretation of it, to your own mediations.

DEBATING WOMEN?

As I walked to the school that morning, I met up with several of my young neighbors who were heading to school for their first classes. One of them, Xiomara, had also been among the group of students who had come to me for assistance in preparing for the debate. As we walked toward the schoolyard I asked if she was ready for her class. "I think so," she answered hesitantly, "although I've never participated in a debate before, and I don't have any idea about what's going to happen." She paused, then added that neither did any of her friends. I tried to comfort her by pointing out that at least that meant they would all be facing the same situation. She smiled, but my comment did not seem to assuage her fears.

While I was sympathetic to Xiomara's uneasiness about the debate, it added to my own curiosity. This would be the first time in my eight months visiting classes at Liceo Parra that I would attend a debate. Indeed, it was the first time I had heard the term used as a specific classroom activity or pedagogic practice. Most of Liceo Parra's social science and humanities teachers employed three teaching techniques: dictation, investigation in groups, and individual research with class presentations. Of these three activities dictation, the most didactic, took up by far the most classroom time. (Science classes also included laboratory time; in both foreign language and mathematics classes, rote learning through memorization predominated.) The other classroom activity that occupied considerable class time was evaluation: exams and their review, pop quizzes, individual reports, and group presentations. There was thus something about the contestability of truth implicit in the term "debate" that made it sound quite distinct from the prescriptive orientation of most teaching I had observed thus far, and rather at odds with the implicit epistemological absolutism of most pedagogy.

No less significant than the novelty of the debate's apparent pedagogic form was the novelty of its content. For this was also the first time I would attend a class in which "woman" (*la mujer*) was the explicit topic of study. Aside from a lesson in the fourth-year literature curriculum devoted to the Venezuelan novelist Teresa de la Parra, and a brief reference to Frida Kahlo by the art history teacher during a unit on Latin American art, individual women were invisible to literature, history, the arts, and sciences. Moreover, women as a group or class of people were never constituted as an object of study. The family was studied in family education and citizenship class; youth and adolescents were studied in psychology class. But there was no place for women, or Woman, in the standardized, nationally certified curriculum.

And so I expected the debate would be no conventional classroom lesson, but rather a unique pedagogical moment whose content disrupted curricular conventions and whose method disrupted pedagogical conventions. However, it is not the debate's two-fold novelty per se that makes it compelling. After all, there is only so much we can learn about "learning" as culturally constitutive social practice by focusing on the spatial arrangements and tangible materials that were offered to or imposed on the students by their teachers or ritualized by the school. Whatever we make of the explicit discourses and even hidden curricula of the school, and whatever the articulated or unconscious aims of the teachers, once the texts, discourses, rules, and other cultural material of the classroom are released to the students, that "production" faces the "ruses" and "fragmentation" of a different, quiet, almost invisible kind of production called "consumption," which is manifest, not in its own products, but "in an art of using those [products] imposed on it" (de Certeau 1984, 31).

Learning, I would argue, is exactly this kind of productive activity—a kind of consumption or use. Much in the spirit of Michel de Certeau's investigation into "the ways in which users—commonly assumed to be passive and guided by established rules—operate" (1984, xi), the focus of my inquiry is on how the students responded to and reconstituted the debate. Like de Certeau, I am interested in "modes of operation" rather than "the subjects (or persons) who are their authors or vehicles" (33). For it is there, outside individual minds or bodies and through the social relations among persons making use of the ideas, images, and materials made available to them, that gender as an ongoing historical project, as a dimension of social inequality *and* individual subjectivity, is substantiated and confirmed but also debated and transformed.

However, where de Certeau's tendency to emphasize the subversive dimensions of use can border on the romantic, the production of new meanings by users may just as likely resort to the proprieties of a particular place and reassert dominant ideologies. While Ms. Martínez may have intended the debate to liberate the students from the confines of conventional pedagogy and curriculum (and the lessons of male dominance that material represented), as the students prepared for and engaged in the debate, they collaborated to create an event that was pedagogically familiar and spatially comfortable, and to make the topic substantively legitimate. I do not think they did this on purpose. I think they did this because the places in which they engaged in these activities insinuated certain "modes of use" and "ways of operating" (de Certeau 1984) that delineated what kind of learning was possible.

Eschewing the romance of resistance, however, does not mean succumbing to the fatalism of social reproduction; and it is certainly not my intention to reduce learning to one or the other pole in a conceptual binarism. Rather, I seek in my account to sustain the complexity and creativity of learning as historical, situated, and social activity that is thoroughly uneven, if not altogether unpredictable, in its meanings and effects.

LEARNING PLACES

The morning of the debate I finished my coffee and *empanada,* put my notebook and pen into my bag, and strolled the short distance along the walkway that passed in front of my house to Liceo Parra. The house I rented as well as the school were part of a residential extension built in the 1960s and 1970s on the outskirts of town. Historically, Timotes's schools had always been located in town, never far from the central plaza (Espinoza Marín 1992). It was thus striking that the present secondary and primary schools had been built on the edge of town, away from Timotes's traditional political, commercial, and religious center. This relocation seemed to index a national shift in the orientation of educational policy that was among the civic transformations initiated with the establishment of a democratic Venezuelan state in 1945 and brought to fruition in the 1960s: a shift away from the socialization of a cultured, local elite (who would, in the case of Timotense educated men, have expected to take on leadership roles in their own town) and toward the preparation of the masses to be, not just educators and professionals, but also technical workers and service providers for a modernizing society (Prieto Figueroa 1951, 1990 [1977]; CERPE 1982).

That this change in the official ideology of formal education was thoroughly gendered was evidenced by the integration of what were formerly all boys' and all girls' elementary schools into coeducational schools. No longer was formal education meant to educate the sons of elite men to become community leaders, while their daughters learned to support them in their charges. The modern Venezuelan state's egalitarian educational ideals posited a genderless, classless student who entered a meritocratic system offering the same opportunities to all.[6] In Timotes this transition began in 1952 when the new (and current) national elementary school merged with the local school for girls, establishing itself officially as *la Escuela Mixta de Timotes* (the coeducational school of Timotes [Espinoza Marín 1992, 66–68]). At the time of my stay in Timotes, every class that Liceo Parra offered was coeducational, including handicrafts and agriculture. Only during physical education classes were students separated by sex to engage in different physical activities.

Coeducation was thus among the spatial practices constituting Liceo Parra as a place meant to prepare all youth equally for the adult occupations they would choose based on intellectual aptitude and civic disposition (Bisseret 1979, 17–27). The pedagogic didacticism of the Timotense classroom ironically reinforced the contradictions of this egalitarian meritocracy: While affirming the teacher's power and objectivity to distinguish individual aptitudes, it minimized the potential for certain (classes of) students to assert themselves or be differentially treated by teachers in any of the systematic, if unconscious, ways that often take place in more participatory classroom settings (e.g., Swann 1989). In most of the classroom activities students were expected to be equally passive receptacles of dictated information; and in most all evaluative activities, students were expected to be equally unimaginative in their regurgitation of memorized information.

While the demure Timotense students generally conformed to these classroom norms with practiced uniformity, they didn't completely acquiesce to the ethos of equality that educational arrangements were meant to promote. Separating themselves into same-sex groupings in classroom after classroom, often despite their teachers' attempts to integrate them, the students sustained the sexual dichotomization learned in elementary school that was more consistent with local gender ideologies and social/spatial arrangements. They similarly defied norms of individualized learning and responsibility by working collaboratively when the teacher wasn't looking, passing school supplies around, and whispering missed words from a dictation to their neighbors. It was striking that, by their third year of secondary school, students engaged in these collaborative ruses in same-sex groupings—not because of any encroaching adolescent shyness, but because the girls had learned that the boys were apt to take advantage of their greater diligence and were reluctant to share their work with the boys.

Officially, the secondary school was located along an educational trajectory from primary school to the university. This trajectory was meant to guide youth from the protective but restrictive confines of the private sphere of family and affectivity into the expansive public sphere of work, politics, and culture—a life course that children supposedly passed along as they were transformed into adult citizens. Meanwhile, through such notions as "aptitude" and "disposition," the egalitarian ideology of the meritocracy naturalized the limited access that female and poor students had to such boundless opportunities as "deficiencies," further constraining female students by constructing educated adulthood as a liberation from the place that was proper to them as women, namely the home (Bisseret 1979).

At Liceo Parra, however, the meritocracy played itself out in unusual ways. As I mentioned earlier, female students generally did better in their classes and stayed longer in school than their male counterparts. Teachers and parents told me that girls performed better in school because their "natural" docility, obedience, and responsibility (the same qualities that disposed them to being contained and protected in the house) disposed them to being good students. Boys were just too unruly to conform to classroom norms and educational expectations. The persistent strength of the local farming economy, along with the enduring status of agricultural labor among Timotenses, continued to draw young men out of school and right to work. Meanwhile, men's labor relieved girls and young women in many families of the pressure to bring in extra household income, freeing them up to continue their studies—at least through secondary school.

But the social arrangements of the school could not escape, transcend, or exist apart from local gendered social relations or the spatial practices through which they were structured and signified; and though six times as many girls might graduate from Liceo Parra as boys, this accomplishment rarely sent them on to the university so as to take on roles as educated women in society. In Timotes the patriarchal family unit organized social and productive relations; and the social and sexual, productive and reproductive relations of the family were structured by spatial practices organized around and signified by the gendered dichotomy of street and house (see Montoya, this volume; Hurtig 1998a). Irreducible to the modern bourgeois gendered division of social life into domestic and public spheres, street and house were distinct "spheres of signification" (da Matta 1991). As a gendered and sexualized spatial configuration it was framed by distinct norms for women's and men's behaviors and mobility that reinforced dominant gender inequalities: complementary and empowering from an androcentric point of view, but contradictory, restraining, and fragmenting from a woman's point of view (see Pratt 1998, 24).

As the expression of networks of familial and productive relations, this spatialized dichotomy had different implications for women and men. The normative spatial and symbolic horizons of women's lives were delimited by the confines of the house (or the farm), extending out in a restricted manner through a network of connected houses of family members and, to a lesser extent, other women friends; men, on the other hand were expected to extend and expand their vistas out and away from their maternal house into the fields or town to labor, into the street to recreate, and out beyond the confines of the town to work, travel, serve in the military, or otherwise engage in those networks of familial and extrafamilial relations that constituted national life.

At the time of my stay in Timotes, the commodified freedom of the street had come to take on increasing value for men, above and beyond the familial responsibilities of the house. Unlike "classic" patriarchy, in which gender inequalities are based on norms of honor, responsibility, and reciprocal familial obligations (Kandiyoti 1991); or autocratic patriarchal relations (which have come to be associated stereotypically with colonial and neocolonial Latin America) based on an arbitrary authoritarianism combining "professed paternalism and calculated exploitation" (Stern 1995, 12), in Timotes male authority and female submission were increasingly sustained through the tension between the expectation of patriarchal responsibility and the legitimacy of male neglect—constituting a gender hegemony I refer to as "negligent patriarchy."

It was the daily arrangements of negligent patriarchy that contributed to the skepticism of the female students when they would hear male students propose that when they got married they would want their wives to work too, and of course they would help with the household chores. "Sure," one young woman responded to this fantasy. "That's what my sister's boyfriend told her. But once they were married he wouldn't let her out of the house." "Not my sister," exclaimed another. "Her husband lets her work, but that's so that he can spend the money drinking and playing the lottery. She works and then is stuck home with the kids." And it was this tension between the desire for spousal and paternal protection and the expectation of neglect—the tension constituting negligent patriarchy—that led female students to declare that they wanted to wait to have a family until they had finished their studies, because, as one student put it, "one needs a career to defend yourself if/when your husband leaves you."

LOOKING FOR WOMEN

Liceo Parra had a small, dark, virtually unused school library with a collection of about 100 textbooks that had remained from previous years. When students had to do research for class presentations they ventured through the street to the town library. Unlike other public places such as the culture center—a refurbished colonial building located off the central plaza with programs and classes frequented by women, men, youth, and children alike—the library was not integrated into the town as a communal space. Hidden down a cobblestone side street, the library was a peculiar spatial extension of schooling into the town itself. Its primary patrons were students and the bulk of its books, many of them textbooks, related to the secondary school curriculum: world history, Venezuelan history, literature, sciences, social sciences, psychology. The main room also boasted an extensive reference collection, including a general encyclopedia and encyclopedias on specialized subjects such as medicine and agriculture. The small side room dedicated to children's books resembled a day care, a reminder of the ideological link between primary education and child socialization.

Students went to the library in small groups after school and occasionally on Saturdays. They would plan these outings carefully, packing extra pens and paper in anticipation of the meticulous copying of lengthy texts that comprised their research activities. More than once students described their forays to the library as preparation for university life, while others said it made them feel like they were *really* students, or they were *really* studying. Distinct from all other activities that took *liceo* youth into town—to the parish for Sunday mass, to the grocery store on errands, to the stadium for a soccer game, to a crafts class at the community center, to a relative's house, or to the post office to call an out-of-town or overseas friend or relation—visits to the library projected students beyond the familial and productive nexus of street/house and into "society" in the abstract as educated persons, extending their imagined adult selves beyond local life.

The library's collection ranged from the mundane to the arcane. Its literature section included Latin American classics and translations of Mark Twain's *Huckleberry Finn*. Its social sciences section included books on the history of Western philosophy, a subject never studied by secondary students. According to the librarian, most of the books they received were donated by the public library in Mérida, the state capital. The Mérida librarians had no idea about how decisions about which books to donate were made; but the presence of such books as *En Defensa del Aborto en Venezuela* (In Defense of Abortion in Venezuela) written by Venezuelan feminist scholar Giovanna Machado (1979), and a Spanish translation of Simone de Beauvoir's feminist classic *The Second Sex* on the shelves of the Timotes library and *not* the Mérida library, made me wonder. Those books and a few others addressing women's issues were clustered together in the Social Sciences section of the library shelves. And yet somehow they were invisible to Liceo Parra students. Indeed, what prompted the students preparing for the debate to come to me in the first place was their frustration over not being able to find any information in the library on Venezuelan women. (One female student from the other team told me after the debate was over that "The hardest thing about the debate was the preparation. I didn't know where to find information about Venezuelan women.")

In their initial perusal of library textbooks and the encyclopedia—the sources they were accustomed to using—they had found one reference to Manuela Saenz, Simón Bolivar's intrepid lover (who was not, strictly speaking, Venezuelan); one reference to the author Teresa de la Parra; and one reference to the contemporary poet Gloria Stolk. And they had done their homework well. Reviewing a study guide on contemporary Venezuelan history prepared for students in the first year of the Diversified Cycle (the fourth and penultimate year of secondary school), the students scrutinized a 26-page listing of "outstanding personalities of national life" entitled "Who has made History?" and found references to five women (Domínguez and Fransceschi 1988).

Not only were the students unable to locate "important" Venezuelan women to determine their "role in society," they also were unable to find information on Venezuelan women more generally, let alone on women's collective social action. For instance, in a textbook on contemporary Venezuelan history, the acquisition of the vote by women was not presented as the successful outcome of a feminist movement to gain suffrage. Instead, it was obscured within a generic description of the 1947 Constitution, which "established the vote as *universal, direct and secret* for Venezuelans over 18 years of age, men or women, literate or not" (Franceschi and Domínguez 1988, 128; emphasis in original). In another textbook the word "women" was not even included in the list of new recipients of this civic privilege. And so women were doubly invisible: invisible as historical actors (whether as celebrated individuals or as a social class) and even invisible as the passive beneficiaries of the historic deeds of famous men. A state of affairs that prompts yet another displacement and reconfiguration of the study of learning practices—from their location within a sociology of knowledge into something Shulamit Reinharz calls a "sociology of lack of knowledge... which studies how and why knowledge is not produced, is obliterated, or is not incorporated into a canon" (1992, 248; see Reinharz and Stone, eds. 1992).

It was a state of affairs that prompted me to do two things with the students I was assisting. First, I arranged to meet them at the library in order to point them to the section on women and sexuality. Several of the students confessed they hadn't known they could use books other than texts and references for their schoolwork. Putting aside the books on women, sexuality, or feminism in general, they selected from the shelf a book that was specifically about Venezuelan women: *Una vida una lucha,* Eumelia Hernández's "summary of events in which the Venezuelan woman has played an important role from her position as citizen, worker and mother" (Hernández 1985, 12). When I asked why they were not checking out the book on abortion, the boys pretended not to hear, while the girls looked down to the floor in that shy and embarrassed way that young Timotense women could so easily adopt, until one of the girls reminded me matter-of-factly that "abortion is a sin."

The second thing I did was provide the students with some of my own materials on Venezuelan women: articles and newspaper clippings on women and education, work, health, and politics, on women's centers around the country, and on International Women's Day celebrations; chapters from *Venezuela: Una bibliografía inacabada,* a compilation addressing "women's participation in national development, 1936–1983" (Ministerio de Estado 1983); and material from the 1989 Venezuelan National Conference on Women. I watched them peruse the material, hoping this fount of information on Venezuelan women's roles in society would somehow impress them, perhaps lead them to contemplate why they hadn't found comparable information in any of their textbooks. Whatever impression it made, they didn't let on. But then Timotense youth weren't generally inclined to proffer their opinions to adults, let alone teachers. (But then they weren't often encouraged to do so, either in the classroom or at home.) The only reaction they shared with me was that it was an *awful lot* of material. For some this meant it would help them do well in the debate; for others it meant a lot of work.

After some negotiation in which the boys tried to foist the work onto the girls while the girls quietly but steadfastly refused, it was decided that they would share the materials with other students on their team so that everyone could "copy the important points" from the readings. Drawing on what they remembered from Ms. Martínez's instructions, they then appointed Alvaro, the most charismatic of the group, to be their director. When Xiomara suggested he take the work they'd done and draw up a list of questions to pose to the other team, he quickly refused, claiming it was too much work for one person to do and suggesting instead that each of them draw up questions from their notes and give them to him. With that decided I left them to their copying.

The following Monday the group of eight dragged a few more of their teammates to my house for our final meeting. Once again they were well-prepared: Each student had diligently copied what she or he believed were the important points from the material they had reviewed. Their selections were striking: From a newspaper article on Venezuelan single mothers, the students copied down percentages of those working, studying, or staying at home and numbers of children born to young mothers; but they passed over the comment that, "after confronting her abandonment by her companion, the Venezuelan mother has to fight against discrimination that she continues to face in the labor market" (*El Nacional* 1991, C8). From a journal article describing contemporary Venezuelan women's political participation, Alvaro copied down percentages of judges and magistrates who were women, and percentages of women elected to legislative positions in 1989; but he ignored the commentary noting that women still do not have a significant presence at the level of "top leadership positions in which fundamental political decisions are made; leaving them primarily in intermediate or technical positions" (Rosillo 1992, 42–43).

If we take learning to be "the construction of present versions of past experience for several persons acting together" (Lave 1993, 8), we can think of the students as drawing on their past experience—of what constituted legitimate information for the classroom, of what kinds of information could be performed back to the teacher for a good evaluation, of how women count and are accounted for in school and in society—to select material they considered appropriate for the debate. This information—numbers, dates, laws, or summary declarations—turned out to be the *least* debatable. The passages they excerpted reduced women's individual and collective struggles and accomplishments to dates and numbers, and rendered their societal roles as possibilities opened to them by men's accomplishments or as assigned statuses. Women had roles in society, changing roles no less; but the conditions of those roles were never of their own making.

DEBATING WOMEN

On the Wednesday of the debate, as on most days, I walked through the main gate, which was casually supervised by a security guard, and into the school yard. During school hours the concrete yard was used for Physical Education classes. Before and after school, the students used this space in ways that both integrated and distinguished the informal recreational practices of the street with the formal recreational practices of the school, applying their own norms to establish which extracurricular activities were coed and which were not. Along the walkway separating the ball courts from the school building, groups of boys could always be found playing handball— a game I never saw girls play. And while male and female students would play volleyball or shoot baskets together, when students used the space to sit and chat they almost always formed single-sex groups. Groups that combined female and male students had one of two purposes: either they were involved in a class project together or they were chatting in clustered pairs of *novios* (girlfriends and boyfriends). As with the seating arrangements of the classroom, the students created learning spaces by arranging themselves rather ritually according to their own gendered terms.

I walked through the next gate, across a patio that gave the school's interior the feel of a colonial house, and up to the second floor in search of Ms. Martínez. When I entered the classroom I was disoriented at first by the strange configuration of the desks, and it took me a moment to locate Ms. Martínez in the classroom. She was sitting, not in her usual place at the front of the room, but on the side against the wall. She signaled for me to take the chair beside her. Except for a nervous shifting of feet, the classroom was silent. The teacher then called on Alvaro to begin the debate. All eyes— those of the 23 students as well as the teacher and myself—fixed themselves on Alvaro. Meanwhile, Alvaro was looking down at a piece of paper clenched tightly in his hands. The trembling paper belied the nervousness behind his staid demeanor. He glanced up at the students across the room then back down at his paper, pondering it with distracted concentration.

Was it the silent attention he commanded that made it difficult for Alvaro to concentrate? Or was it the discomfort he felt sitting in a classroom that had been completely rearranged for this specific event: the desks configured in an unfamiliar way, creating a spatial relationship among the students and the teacher that was new and strange? After all, day after day for the past nine years, Alvaro had entered the classroom and taken his seat with the other students at desks lined up in fairly neat rows all facing the teacher and the blackboard at the front of the room. And for the past six months, Alvaro had entered this particular classroom, headed straight to the right side of the room toward the back, and set himself down amongst the other nine male students who inevitably clustered themselves apart from the fourteen female students. This habitual seating arrangement was only disrupted when a teacher assigned a research project and divided the students into study groups. On those days the students would arrange their desks in groups of four or five in order to work together. Since the girls preferred not to study with the boys—"they are lazy and leave us to do all the work," commented the female students—Alvaro rarely found himself sitting in a group with girls during class time.

But now, as Alvaro pondered that vital piece of paper, he found himself sitting on one side of the room with half the class, in the first of four rows of desks that were lined up facing the other half of the class, which was similarly arranged. Moreover, there were girls sitting on either side of and behind him, while his three male teammates were sitting together off to one side. And if the novelty of all this weren't enough to make him fidgety and uncomfortable, he had just been called upon to initiate an educational activity that neither he nor any of his classmates had ever participated in before nor seen enacted in other classrooms. In staging the debate, not only were the students divided into teams and pitched against each other in a sequence of inquisitional turn-taking that was entirely novel, but also the traditional locus of knowledge and authority—the teacher—remained on the spatial and performative margins of the event. The debate thus had an unfamiliar pedagogic form and uncertain evaluative standards that resisted easy assimilation to other more familiar practices. No wonder Alvaro was nervous!

The near-silent seconds passed. Then Alvaro, lifting his head slowly and deliberately, fixing his gaze at the students in front of him and artfully imitating the gestures of a teacher about to interrogate the class, finally spoke: "Name two laws that benefit the Venezuelan woman of today, and the dates of their enactment." In his enunciation, his entire self seemed to embody pedagogic authority. But it was only practice. Without turning his head, Alvaro's eyes darted toward the teacher, searching for a nod, a raised eyebrow, any gesture of approval. But the teacher, her arms folded, remained impassive. She turned her head to the opposing team and awaited their response.

With his boldly enunciated and succinctly stated question, Alvaro had simultaneously initiated his first debate and constituted "Venezuelan women" as a legitimate topic of study through which his performance as a student could be evaluated. Alvaro's question was followed by silence. The members of the opposing team looked at each other, then at the teacher, then to their director. Timid hands went up then back down; the students kept glancing over to the teacher with looks of mild distress. Apparently the students had no idea who was meant to respond, and the certainty of the answer's content was quickly overwhelmed by the ambiguity of the activity's form. Determining Venezuelan women's role in society was creating a panic!

Ms. Martínez, impassive as ever, identified the problem and clarified the uncertainty: The person asking the question could request the answer from one person, or open it up to the entire team. All eyes darted back to Alvaro who, taken aback by his additional power, fumbled a bit and then called on one of the female students across the room. "The reform of the 1982 Civil Code," she promptly responded. He glanced down at his papers and then back up to the group. "That is correct. And the second?" he added more confidently, apparently beginning to enjoy his place at center stage.

The young woman looked down at her paper, glanced around to her teammates to see whether one of them had raised their hand, to her friend beside her who was motioning to her paper, and then to Ms. Martínez to see whether she could confer with her classmate. Since the teacher said nothing, and since students were in the habit of conferring with each other and comparing notes on most class work except exams, the student bent over and read from her friend's paper. "Two: the 1982 law against all forms of discrimination against women." Alvaro glanced at his paper, which apparently did not include that datum. He was about to rule the answer incorrect when Xiomara tapped him from behind. "Yes, it is correct," she whispered, showing him her paper. "You have correctly answered the question," he declared. I turned to Ms. Martínez, commenting that the student hadn't indicated what the reformed civil code had done for women. (It eliminated all legal impediments for married women to acquire property in urban areas.) She responded that this was Alvaro's fault, since he hadn't asked. After all, the student *did* answer the question as it was posed. I asked her whether the students were aware of that. She was sure they were, since those were the terms by which their responses to exam questions were evaluated.

The opposing team then reviewed their list of questions together and their director, also a male student, responded with a "date" question: "In what year did Venezuelan women first exercise their right to vote?" he asked, invoking the vocabulary and syntax of the same textbooks that had omitted this very information. Alvaro's team spontaneously conferred among each other, affirming cooperation as a legitimate debate practice. One of the girls raised her hand just as Alvaro replied: "Venezuelan women first exercised their right to vote in 1946." "Yes, it is correct," replied the interrogator.

And so it went. The questions posed were of three general kinds. There were empirical questions about dates marking changes in women's juridical status or political events, for instance. Then there were descriptive questions posed in such a way as to require short answers, as though the list of possibilities were finite, and as though there were no margin for debate: "In what three ways is women's condition better today than it was in the past?"

"What are the most common health problems for modern women?" And there were more expansive questions, similarly presented in such a way as to imply that there was only one correct answer, and that it could apply to all Venezuelan women, as a unitary category: "Is there discrimination against women today, or not?"

As the students carried on the debate, I began to notice that neither team posed questions about the lives or acts of individual Venezuelan women, whether in political or cultural life. Nor did either team ask about women's collective struggles or accomplishments. Because I had given Alvaro's team plenty of material on both subjects, it is possible that this selective omission reflected a collective interpretation of the subject at hand: that debating the "role of Venezuelan women in society" meant considering how women's roles were determined by society, and not vice versa. But we should also wonder whether the near exclusive celebration of famous men in textbooks, and the severe repression of women's collective activity outside the house, made learning about Venezuelan women as historical actors an impossibility for that group of Timotense students.

While the students delineated the parameters of the debate's content in terms that were substantively familiar or comfortable, they also modified the initial procedures for turn-taking that ran counter to the standard classroom hierarchy. About midway through the debate one of the female students (not from Alvaro's team) raised her hand and asked the teacher whether she could pose a question. Ms. Martínez responded that of course she could, since it was never stated that only the directors could ask the questions. (Was I imagining a glimmer of delight on her face in response to this young woman's assertiveness?) The young woman promptly asked, "How was a woman looked upon in the Gomez days,[7] if she abandoned her home to work?" Several hands went up, and the girl posing the question called on a female student to answer. Without looking at her papers she intoned her reply as though in catechism class: "During the dictatorship, a woman who left the home to work was abandoned by her family and had to defend herself alone." The interrogator, appearing rather flustered by this confident response to her question, looked down and mumbled softly, "Yes, that is correct."

This new, more democratic debating practice quickly engendered a division of intellectual labor by sex, in which the male students asked and responded exclusively to quantitative questions or questions regarding dates of events, leaving it to the female students to pose and address the more descriptive questions. Through their deliberations on Venezuelan women in the past and present, the female students colluded in the construction of an insistently positive account of the changes in Venezuelan women's role in society. Girls from both teams asked questions about women's status in the past or asked for comparisons between women's role or situation in the past and today. (Curiously, the terms "role" and "situation" came to be used interchangeably as the debate went on.) In every aspect of social life that was considered, women's situation today was portrayed as better than in the past. In a sense, then, this collective account of women's changing role in society was not so much progressive as it was binary: In the past women were kept in the home, but now they can work in many fields; in the past men were allowed to beat their wives, but now they are not; in the past there was discrimination against women, but now there is not.

At the same time, through their referencing of the subject of their debate, the male and female students positioned themselves differently in relation to that subject and each other. The male students used only the third person, "she" or "they" but never "you"—not daring, or perhaps refusing, to include their female companions within the objectified category of "Venezuelan women." If generic women could be the subject of debate, of formal education, and thus of history, the women they knew could not.

Where the male students were uniform in their referencing of "Venezuelan women in society," the female students were not. But the young women's references had their own consistency that implicated them in their binary historical account of Venezuelan women's role in a particular way. When they discussed the role of Venezuelan women in the past, the female students had recourse to the third person singular—"she." In this way Venezuelan women became a unitary other when presented as the victim of a past *machista* barbarism—a past from which female students wished to distinguish their own situations and possibilities, as well as the state of their nation.

Where the female students were consistent in distinguishing themselves from Venezuelan women of the past, they were less consistent in their referencing of contemporary women—alternating ambivalently between "they" and "we" as if to index their uncertain efforts to construct themselves as modern women in-the-making. Whether they felt too young or too modest to think of themselves as Venezuelan *women;* whether they were uncertain whether the optimistic portrait they had created pertained to them, I can't say. But the uses of pronouns would seem to suggest whose role in society, for the students, was and wasn't subject to debate.

RESOLVING WOMEN

If the young women in Ms. Martínez's class were reluctant to identify themselves with Venezuelan women in the past, they nonetheless rendered the history of Venezuelan women's role in society in celebratory terms that acknowledged women's past sufferings, and perhaps implicitly validated women's struggles. At the same time, the modern-day scenario these young women colluded in creating was at odds with the everyday realities they regularly discussed amongst themselves and even with me in our discussions: realities faced by their mothers, aunts, sisters, and themselves. What happened to their personal historical understanding of the value of a career as a tool "to defend oneself when your husband leaves you"? What happened to their ongoing identification of the repressive practices and beliefs of their fathers, brothers, and boyfriends? Subtly, stupendously, the female students had gradually occupied the performative center of the debate, appropriating the process of constructing Venezuelan women as their own. And yet, through the spatial practices of the debate these young women seemed to have created yet one more dream of hopeful unreality.

Ten minutes before class was to end, Ms. Martínez's mild voice cut gently and abruptly through the bustle of the debate in full gear. "Team 1, prepare your last question." As the students huddled together to decide on a question, Ms. Martínez explained to me how she was going to determine the outcome of the debate: "I will give each team points for the quality of each question and the accuracy of each answer." As the students waited with anticipation, their gazes and bodies shifted toward Ms. Martínez and the familiar configuration of knowledge and authority was quickly reestablished through the practice of evaluation. "Well," announced Ms. Martínez, almost cracking a smile. "I have to congratulate both teams: Team 1 got the most points for questions, but Team 2 got more points for answers."

"You mean there is no winner, Profesora?" blurted out one of the male students, appearing rather crestfallen.

"No, I mean you have carried out a very good debate." Ms. Martínez smiled. "Class is dismissed."

NOTES

Thanks to Hal Adams, Rosario Montoya, Lisa Rosen, and Lessie Jo Frazier for comments on earlier versions of this chapter. Thanks also to participants in the 2001 Spencer Foundation Advanced Studies Seminar on Anthropology and Education for inspiring me to rethink education in all its dimensions.

1. Fieldwork from 1990 to 1993 was supported by the Woodrow Wilson Foundation, the Organization of American States, and the University of Michigan.
2. All names except the town of Timotes have been changed.
3. In general secondary school teachers are called *profesor* or *profesora*, while elementary school teachers are referred to as *maestro* or *maestra* (teacher).

4. All translations of spoken and written texts are mine.
5. Luykx (1997), Levinson (2001), Gill (1997), and Stromquist (1997) are examples of such works that address issues of gender.
6. The central tenets of the Venezuelan Teacher State (*Estado Docente*), first stated in the 1947 constitution, designated education as an essential function of the State, and every person as entitled to study sciences or arts, within the limitations of curriculum and guidance established by the State (Prieto Figueroa 1990 [1977], 87–89).
7. Juan Vicente Gómez was dictator of Venezuela from 1908 to 1936.

REFERENCES

Bisseret, Noëlle. 1979. *Education, Class Language and Ideology.* London: Routledge and Kegan Paul.

CERPE. 1982. *El sistema educativo en el proceso de modernización de Venezuela.* Colección CERPE no. 3.

De Certeau, Michel. 1984. *The Practice of Everyday Life.* Trans. Steven Randell. Berkeley: University of California Press.

Domínguez, Freddy and Napoleón Franceschi. 1988. *Trabajos prácticos de historia de Venezuela contemporánea.* Caracas: Colegial Bolivariana, C.A.

El Nacional. 1991. "Cuando las madres son también padres." May 14.

Espinoza Marín, Jesús María. 1992. *Historia mínima de Timotes.* Timotes-Mérida, Venezuela: Centro Editorial Escuela de Comunicadores Populares "Mario Kaplún."

Franceschi, Napoleón and Freddy Domínguez. 1988. *Historia de Venezuela. Octavo Grado.* Caracas: Colegial Bolivariana, C.A.

Gill, Lesley. 1997. "Creating Citizens, Making Men: The Military and Masculinity in Bolivia." *Cultural Anthropology* 12(4): 527–550.

Hernández, Eumelia. 1985. *Una Vida Una Lucha.* Caracas: Fundación para el desarrollo social de la región capital.

Hurtig, Janise. 1998a. "Gender Lessons: Schooling and the Reproduction of Patriarchy in a Venezuelan Town." Ph.D. diss., University of Michigan.

———. 1998b. "Myths of (fe)male achievement." *La Educación, Interamerican Review of Educational Development* 42: 101–120.

Kandiyoti, Deniz. 1991. "Bargaining with patriarchy." In *The Social Construction of Gender,* edited by Judith Lorber and Susan Farrell. London: Sage Publications.

Lave, Jean. 1993. "The practice of learning." In *Understanding Practice: Perspectives on Activity and Context,* edited by Seth Chaiklin and Jean Lave. Cambridge: Cambridge University Press.

Levinson, Bradley. 2001. *We are all Equal: Student Culture and Identity at a Mexican Secondary School: 1988–1998.* Durham: Duke University Press.

Levinson, Bradley, Douglas E. Foley and Dorothy C. Holland, eds. 1996. *The Cultural Production of the Educated Person.* Albany: SUNY Press.

Luykx, Aurolyn. 1997. "Discriminación sexual y estrategias verbales femeninas en contextos escolares Bolivianos." In *Más allá del silencio: Las fronteras de género en los Andes*, edited by Denise Y. Arnold. La Paz: CIASE/ILCA.

Massey, Doreen. 1994. *Space, Place and Gender.* Minneapolis: University of Minnesota Press.

Matta, Roberto da. 1991. "Espaço: Casa, rua e outro mundo: o caso do Brasil." In *A casa y a Rua.* Rio de Janeiro: Editora Guanabara Koogan S.A.

McDermott, Raymond P. 1993. "The Acquisition of a Child by a Learning Disability." In *Understanding Practice: Perspectives on Activity and Context*, edited by Seth Chaiklin and Jean Lave. Cambridge: Cambridge University Press.

Machado, Giovanna. 1979. *En defensa del aborto en Venezuela.* Caracas: Editorial Ateneo de Caracas.

Ministerio de Estado para la Participación de la Mujer en el Desarrollo. 1983. *Venezuela: Bibliografía inacabada. Evolución Social 1936–1983.* Caracas: Ediciones Ministerio de Estado para la Participación de la Mujer en el Desarrollo y Banco Central de Venezuela.

Montoya de Solar, Rosario. 1995. "Fractured Solidarities: Utopian Projects and Local Hegemonies in Nicaragua, 1979–1990." Ph.D. diss., University of Michigan.

Poovey, Mary 1988. *Uneven Developments: The Ideological Work of Gender in Mid-Victorian England.* Chicago: University of Chicago Press.

Pratt, Geraldine. 1998. "Geographic Metaphors in Feminist Theory." In *Making Worlds: Gender, Metaphors, Materiality*, edited by Susan Hardy Aiken, Ann Brighma, Sallie A. Marston, and Penny Waterstone. Tucson: University of Arizona Press.

Prieto Figueroa, Luis Beltrán. 1951. *De una educación de castas a una educación de masas.* La Habana: Editorial Lex.

———. 1990 [1977]. *El Estado y la educación en América Latina.* Caracas: Monte Avila Editores.

Reinharz, Shulamit. 1992. *Feminist Methods in Social Research.* New York: Oxford University Press.

Reinharz, Shulamit and Ellen Stone, eds. 1992. *Looking at Invisible Women: An Exercise in Feminist Pedagogy.* Washington, D.C.: University Press of America.

Rosenblat, Angel. 1964. *La educación en Venezuela.* Caracas: Monte Avila Editores.

Rosillo, Carmen. 1992. "La participación política de las mujeres en Venezuela." In *Fermentum: Revista Venezolana de Sociología y Antropología* 2(4): 37–51.

Stern, Steve. 1996. *The Secret History of Gender: Women, Men, and Power in Late Colonial Mexico.* Chapel Hill, N.C.: University of North Carolina Press.

Stromquist, Nelly. 1997. *Literacy for Citizenship: Gender and Grassroots Dynamics in Brazil.* Albany, NY: SUNY Press.

Swann, Joan, 1989. "Talk Control: An Illustration from the Classroom of Problems in Analyzing Male Dominance of Conversation." In *Women in their Speech Communities*, edited by Jennifer Coates and Deborah Cameron. London: Longman.

Taussig, Michael. 1992. *The Nervous System.* New York: Routledge and Kegan Paul.

2. "To Act Like a Man" ᘓ

Masculinity, Resistance, and Authority in the Ecuadorian Andes

Barry J. Lyons

INTRODUCTION

This chapter focuses on notions of masculinity and resistance among Quichua-speaking indigenous villagers, or *Runa*, in Pangor, a parish in Chimborazo province in the central highlands of Ecuador. Indigenous Pangoreños' notions of masculinity have been intertwined with authority and resistance in class, ethnic, and age-based hierarchies. This essay deals primarily with the early to mid-twentieth century, when haciendas dominated the area, but also examines the contemporary period of indigenous ethnic mobilization.

An explicitly gendered notion of insubordination reflected and reinforced an association of resistance with wildness. This association itself reflected a social and symbolic spatial order in which people lived and grew their crops on hacienda land and village elders were linked to hacienda authority. While hacienda residents' partially autonomous social and economic life helped sustain a critical vision toward the hacienda, they lacked a sufficiently autonomous social space that could serve as a base for collectively articulating an alternative to the hacienda system. Hacienda residents seem to have understood resistance more as an individual act of wild masculine self-assertion than as a sign pointing to a possible new society.

This resistance—limited and unorganized as it often was—did nonetheless contribute to the demise of the hacienda system (Thurner 1993). In the post-hacienda period, indigenous villagers have collectively gained formal local autonomy; the village has changed from a site of ethnic and class domination to become a site of indigenous ethnic resurgence. Pangoreños have joined in a growing, organized ethnic political movement and, in the process, they have been reworking the associations between resistance, authority, society, and the wild. This reworking underlines the dialectical relationship between forms of resistance that "tear down" boundaries, on the one hand, and the counter-hegemonic reconstruction of local social order, on the other.

In examining local discourses on gender, authority, and resistance ethnographically, one of my aims is to suggest an avenue for critically rethinking the concept and the popularity of "resistance" in recent academic discourse. Several scholars have questioned the academic "romance of resistance" in terms of the ways it tends to oversimplify social and cultural processes.[1] A close reading of an unfamiliar discourse on (and of) authority and resistance may ultimately help us to look at our own notions anthropologically by de-familiarizing them in an alien mirror.

I became interested in local notions of masculinity while trying to understand how laborers viewed hacienda authority and responded to hacienda oppression. Haciendas belonging to the mestizo elite occupied most of the land in Pangor, as in much of the Andes, until the 1960s or 1970s; some haciendas belonged to institutions such as the Catholic Church, which rented them out to other landlords. Resident indigenous laborers were granted plots to farm, along with access to hacienda pastures for their livestock. In return, landlords required one man from each household to work four or five days each week in agricultural or other labors, while periodically calling on women and children for other services. Inherent in this system was a chronic tension between hacienda residents' need to attend to their own subsistence and the landlord's demands for their labor. The hacienda enforced its demands through coercion: Laborers who missed work or disobeyed orders risked a whipping, and landlords sometimes expelled recalcitrant workers from their estates.

In talking about the hacienda period with my informants—former hacienda laborers, now members of an autonomous village with their own land—I was unable to find a Quichua word that corresponded exactly to what I thought of as "resistance," that is, active self-assertion (covert as well as overt) in the face of power. The Quichua word meaning something similar that I heard most frequently was *cariyana*, built on the root *cari* or "male."

My title, "to act like a man," is a loose translation of *cariyana*. The suffix *-ya* means "to become," both in the sense of a long-term transformation and, as in this case, of momentarily taking on a particular characteristic ("to get" a certain way): "to get male" might be a more precise translation, paralleling, for example, *cushiyana* (to get happy) or *mapayana* (to get dirty).[2] "Cariyana" is used in different ways in different places (cf. Haboud de Ortega et al. 1982, 35; Múgica 1979, 85). In Chimborazo province, or at least in Pangor, however, "cariyana" means "to challenge authority." For example, it is "cariyana" for laborers to have talked back to stewards, or children to their parents. While both females and males can grammatically "cariyana," the lexical link to masculinity is meaningful: Speakers also invoke masculinity in other ways to underline someone's ability or inclination to challenge authority.

These observations suggest several questions about masculinity, authority, and resistance. First, what is the connection between masculinity and insubordination—what is it about masculinity that makes challenging authority a "male" thing to do? Is it a "male" thing in the sense that only males should do it—when a female "gets male" (challenges authority), does she thereby lose some valued aspect of her femininity? What other things do Pangor Runa talk about as markedly "male"? Furthermore, what moral connotations does cariyana have—is it positively valued, like "standing up for oneself like a man" in English or like "resistance" for anthropologists? What place does cariyana—or for that matter, masculinity more generally—have in local conceptions of boys' and men's social roles, and especially what place did it have in the hacienda period?

While my primary concern is to understand the place of cariyana and masculinity in the hacienda period, I only had direct access to views of gender during my fieldwork in the 1990s. How fully and accurately accounts of the hacienda period gathered in the 1990s reflect the gender ideology of the period they describe, some three decades earlier and more, is difficult to assess (see Lyons 2001; Thurner 1993). The strategy pursued here does have certain merits, however, for painting a tentative picture of images of masculinity during the hacienda period itself. First, I do not limit my attention to the explicit content of accounts of the past—the level of meaning most subject to deliberate refashioning and unconscious reinterpretation. The analysis is also based on implicit associations and meanings embedded even in the very form of words and expressions such as "cariyana." Such linguistic forms are relatively stable, as they are not under the conscious control of individual speakers. Along with oral histories, I also rely on narratives that one might broadly call mythical, told by men who reached adulthood and in most cases middle age under the hacienda, and who themselves told of having heard these stories from their elders. While such narratives surely vary with each performance, there's no doubt they manifest oral traditions that predate the last few decades. Ethnographic observations made during the 1990s are helpful in interpreting other sorts of evidence and will offer insights into how the relationship between gender images, authority, and resistance is changing in the post-hacienda context.

Cariyana might be seen as a local form of desalambrar, challenging boundaries and distinctions between the wild and the social, juniors and seniors, Runa laborers and mestizo bosses. Yet the current indigenous political movement—since 1990 the most important force challenging social inequalities in Ecuador and one of the strongest Indian movements in Latin America—has found less of a place for "cariyana" in its rhetoric than for cariyana's traditional semantic opposite, *respeto* or "respect." I indicate toward the end of this chapter some of the ways that the indigenous movement's project of remaking society has entailed reconstruction as well as "tearing down." The indigenous movement has worked to rebuild local society, ethnic identities, community solidarity, and communal authority—a rebuilding that has both fed into and fed off of practices of resistance to the state and mestizo elites.

THE MEANINGS OF CARIYANA

Perhaps the most common form in which my informants use the verb "cariyana" is as a negative imperative, *ama cariyangui* (Don't get male/challenge authority); this is how they recall hacienda authorities and indigenous elders admonishing their subordinates. In this context, "cariyana" can refer to improper, insolent behavior toward parents or other authorities who deserve respect. My informants' accounts and their day-to-day interactions do not manifest any assumption that it is an inherently noble act to question or defy authority—especially not the authority of parents and other elders. Indeed, in ritual admonishments (see Lyons, n.d.), "cariyana" seems almost to be used as an antonym to *respetana*, "to respect," a term that stood high in the hierarchy of moral values and encompassed a variety of religious and social obligations.

Informants do nonetheless also describe themselves as cariyana-ing as a justified response to abusive hacienda authorities. José María Pillajo, for example, tells of an encounter he and two workmates had with a notorious hacienda steward:

> Because he was cruel, ... we cariyana-ed. When we were working up in the mountains and he went to strike, carrying a whip, we cariyana-ed. Plowing, holding the goad for the oxen, ... "Let's see! Give me one, strike me, *carajo* [dammit]!" He stopped there, *carajo*. "Before I hit you with the goad, *carajo*, get out of here." Then he just left, muttering under his breath.

Recounting the same incident in another interview, Taita José comments, "We were *macho*, we didn't support their abuse, *carajo*, we didn't just take it."

Josefa Shagñay also recalls her verbal duels with the indigenous over-seers, noting that "only those of us who were prone to cariyana would cariyana (cariyaclla cariyac canchic). Those who weren't prone to cariyana, no. ... I would cariyana. So they didn't want to see my face."

The fact that a woman like Josefa Shagñay can speak with pride of her male-like rebelliousness is congruent with the generally flexible and relative nature of indigenous Andean gender and other dualisms, which do not operate according to a logic of either/or mutual exclusions (see, e.g., Allen 1988). Thus, while drinking, strength, and defiance of authority are marked as "male," women may also drink, be strong, or cariyana without ceasing to be feminine. Josefa Shagñay did not seem to feel that her rage and defiance toward abusive hacienda authorities and an abusive first husband made her any less of a woman, less attractive to men, or less able to nurture her chil-dren. This contrasts, for example, with Esperanza, a Mexican woman whose life history Ruth Behar has written; Esperanza believed her rage toward her husband poisoned her nursing infants (Behar 1993).

Another way to pursue the relationship between dispositions or actions such as cariyana and male or female gender identity is to inquire into the ways the Quichua language encodes essentialist or nonessentialist under-standings of identity and social action. The issue turns out to be quite com-plex. In discussing themes of gender or defiance, Quichua speakers make regular use both of formulations that suggest a constant identity—"we too [not just the bosses] are male" (ñucanchicpish carimi canchic)—and of the verb, "cariyana," which seems to suggest more of a process, a temporary assumption of manly qualities. Yet this verb appears most commonly in my interviews in a form that can be interpreted as a nominalization, *cariyac*. As an agentive suffix, *-c* corresponds to the English *-er*, so that *cariyac* might be translated as "defier" or "one who defies." In other words, the word seems to point to a type of person, not simply an action. On the other hand, in context, especially as combined with the verb "to be" (*cariyac cani*), *cariyac* can also be interpreted as a verb indicating a habitual, indef-inite past (see Catta Q. 1994, 150, 156, 162–164; Cusihuamán G. 1976, 172–174). I translate this as "I would cariyana" (not "I am a cariyana-er"). What the translation misses is the (essentialist?) sense of identification with a habitual disposition that the Quichua grammatical structure seems to carry. Yet again, in the case of "cariyana," this disposition itself is an incli-nation to be transformed, as encoded in the suffix *-ya*. It seems, then, that Quichua structures can be interpreted in processual or essentialist terms. A more definitive answer would entail a much fuller and more fine-grained analysis of Quichua grammar, lexicon, and actual speech samples than I can attempt here.

Be that as it may, the word "cariyana" does certainly encode an association between masculinity and defiance. Why is it especially "male" to challenge authority? What other "male" qualities does challenging authority involve? Strength is of obvious relevance, inasmuch as a challenge to authority could lead to a physical confrontation, as José María Pillajo describes. Less obvious is an association between maleness and speech, particularly strong, fearless, even offensive speech.

Soon after a baby (male or female) is born in Pangor, the midwife or mother takes a piece of straw and moves it across the infant's lips, symbolically sewing its mouth shut. This is so that the child will not be insolent— bad, offensive words will not come out of its mouth. Rural Ecuadorians frequently lace their speech with diminutives, hedges, and other signs of respect and esteem; negative emotions tend to be expressed indirectly, often through irony.[3] Cariyana breaks with this restraint; perhaps the paradigmatic form of cariyana is *respondina* (to talk back). Josefa Shagñay says of her younger brother, "he too learned to cariyana"; like her, he would "stand up" (*shayariclla*) to the steward and overseer. Both of them gained a reputation among their peers and bosses for speaking bad words—a reputation that she says, along with her temper, persists to this day: " 'You speak badly,' they say to us, 'you are bad people.' ... 'Male and female, both of you have an ugly mouth, a bad mouth.' " Again, though this account comes from a woman, strong, direct speech seems to be viewed as a characteristically male trait. Note the explicit invocation of masculinity in the following discussion of indigenous people's increasing ability to speak up in public and political contexts:

> Now, in these times, it's not like in the old days .... Now ... there are knowledgeable males, knowledgeable females. ... Now, we laborers too, words, words, like a male, like a male, we're speaking [shimi shimi cari cari ña rimarinchic]. I, though a laborer, I speak, *cari cari,* I don't get frightened now. Tomorrow, ... I'm going ... [to a demonstration in Riobamba].

Thus there seems to be a chain of associations between maleness, strong speech, and challenging authority.[4]

Strong anger is a third feature linked to challenging authority that likewise seems to be viewed as characteristically male. In explaining her disposition to cariyana, Josefa Shagñay speaks of an anger that overcomes her and renders her fearless of any consequences. For male subjects, a construction very similar to "cariyana"—*cari tucuna*—is often used to mean something like "to get angry." The verb *tucuna* means something like "to become" or "to make like"; "to get angry" as *cari tucuna* is thus, literally, "to become or make like a male." Strong anger is also linked in an interesting way to wildness: eating the meat of wild animals (especially of bears) is said to make a person prone to feel strong anger. Men take pride in their hunting skills, often posing for photographs with a hunting rifle. Masculine anger, then, may be one example of a broader association between masculinity and wildness.

MASCULINITY AND THE WILD

A Quichua dictionary developed by a team of linguists and Quichua speakers from all over Ecuador illustrates the meaning of "cari" with this sample sentence: *Urcucunapica cari huacracunatalla charinchic* (In the mountains, we have male cattle only) (Haboud de Ortega et al. 1982, 35). The sentence links masculinity to place and wildness. Bulls that graze unconfined in mountain pastures with little human contact become semiwild and fierce, creating great excitement when they are brought down to village plazas for bullfighting during fiestas. Another word incorporating "cari" also suggests a semantic opposition between masculinity and domesticity: Parents may chide a daughter who avoids or bungles a household task as *carishina* (male-like).

Some caution is required in interpreting these associations and oppositions, however. Some years ago, Olivia Harris demonstrated that gendered notions of the wild in the Bolivian Andes are ambiguous, shifting, and extraordinarily complex (1980). The same point applies to Pangor. As elsewhere in the Andes, the primary spaces of wildness in Pangor are the high mountain pastures and peaks above the areas of cultivation and human settlement. These spaces are conceived of as more masculine than feminine in some ways, as the dictionary sentence exemplifies. At least after marriage, men are (and were in hacienda times) somewhat more likely than women to visit these spaces as herders and travelers, or as shamans to work with the mountains. Yet the mountains themselves are not always male. Complementing the *urcu runas*, male "mountain men" sometimes identified with the devil, are *urcu doñas*, "mountain women" likewise identified with the devil. Male cowhands might chance upon an *urcu doña* in the form of a beautiful indigenous woman bathing nude in a high mountain lake.

Indeed, some of the most important local and regional mountains are female. Locally, Aichi and Navag are the sites of two important passes linking the Pangor basin to neighboring basins. Both have been partly Christianized (and perhaps quasidomesticated) as "virgins" or sites of apparition of the Virgin Mary, and they are sites and objects of pilgrimage and fiestas. The two highest mountains in the region are Chimborazo and Tungurahua, respectively identified as husband and wife. "Mama Tungurahua" may be the more important of the two, as her mountain is identified with Purgatory or Hell (see Lyons 1999). Thus, masculinity may be wild, but the wild is not intrinsically masculine. Pangoreños' notions of what it is to be cari seem to entail some sort of engagement with wildness, but wild places and beings may be male or female.

Harris also makes the important point that men and women "change their relationship to the symbolic"—and to wildness—"in the course of their lives" (1980, 92). She notes that for the Laymi in Bolivia, naming rituals, the acquisition of speech, and finally marriage are all steps in the transformation of presocial creatures into fully cultural persons. The married couple embodies society, and therefore, in some representations, unmarried: married is understood as wild: cultural (87–88, 90). Harris's points, again, apply with equal force to Pangor—with one additional step. In the hacienda period, a married couple was only fully incorporated into adult society when the couple sponsored a fiesta, normally within the first few years after marriage. Fiesta sponsorship was said to "wash the face" (*ñahuita maillana*) and earn the couple honor and respect. Hacienda bosses and Runa elders also saw the experience of fiesta sponsorship as taming troublesome young men, making them less inclined to cariyana and more inclined to respect authority. A mature man did not cease to be cari, and men continued to engage with the wild, but the nature of that engagement changed.

In stories and other talk about wild spaces and creatures, males engage in different ways with wildness. Some nonhuman male beings epitomize wild, asocial, or anti-social qualities, and human males can become wild in this sense; sometimes, it is a male role to dominate or domesticate the wild and turn it to social use. I cannot point to a direct link between the term cariyana or authority relations and this talk about the wild. I don't have any stories of hunters, for example, consuming bear meat and then pointing their rifles at the landowner. The *chuzalungu*—a boy-devil who lives high up in wild mountain pastures—does not openly confront human authority; he simply flees from society. I'll also tell of a *vaquero*, a cowhand on a hacienda, who transforms himself into a wild bull. It would make for a simple argument for cariyana as wildness if as a bull he turned on the landowner and gorged him, but that's not how the story goes.[5] If there is a link between cariyana and the masculine engagement with wildness, it's because of the general opposition between the wild and the social, the general link between society and authority, or between the rejection of social norms and defiance of authority.

The chuzalungu figure is often referred to as a devil, sometimes as an *urcu runa*, a sort of personification or owner of a mountain. José Pillajo's grandfather, a hacienda cowhand, saw a chuzalungu next to a high mountain lake,

> mounted ... on a fierce bull. ... Grandfather [quietly] watched. The youth got down from the bull and walked around, he said, touching the bull's testicles, rubbing the horns. ... The bull didn't do anything. It didn't try to gorge him, it didn't get up, nothing. ...
>
> Grandfather yelled out, he said ... [and] the cattle ... all stood up, agitated, like fierce cattle. The boy disappeared into the lake.

The chuzalungu is no longer there, Taita José said, because people brought some llamas to this mountain. The chuzalungu flees from llamas and pigs and prefers quiet areas empty of human presence.

The chuzalungu's most prominent feature is his monstrously long sexual organ. This does not represent any ideal of how human males should be, but rather, represents an amoral figure of Freudian fantasy. The chuzalungu's sexuality is fundamentally antisocial and hence not fertile: In some stories, he rapes young women who initially mistake him for a child, disemboweling and killing them in the act. The chuzalungu is said to pursue single women in particular. In one case, a woman who pastured hacienda sheep in the mountains began to hear a voice calling in the fog; her elders told her this could be the chuzalungu pursuing her, intending to seduce her. Out of fear, maybe worried she could be seduced, she decided to marry the man who had been courting her (who told me the story in 1992, perhaps some five decades later; RS 92/9). The chuzalungu, then, seems to represent a kind of wild masculinity, especially male sexuality; his childlike body with the monstrous sex organ expresses his stunted development, arrested in an asocial phase, symbolically opposed to marriage and everything else domestic—a wild boy forever.

The story of the vaquero or hacienda cowhand who transforms himself into a bull represents a different sort of relationship between masculinity and the wild. If the chuzalungu is a wild, boylike creature, the vaquero is an adult, married man who regresses to a wild, animal-like state.

> There was a vaquero who would go to pasture the cattle in the mountains. He left his house in the morning without eating breakfast.
>
> Up in the mountains, there was a lake. There, that vaquero sounded his horn, and all the cattle gathered together. Then the vaquero took off all his clothes, leaving them by the side of the lake, and entered into the lake. He emerged from the lake on the other side as a bull. ... He then passed the day mounting the cows and fighting with the other bulls.

At the end of the day, he walked into the lake again and re-emerged as a man. He got dressed and returned home.

The vaquero's wife noticed that his face was all scratched up [from fighting with the other bulls]. She asked him what had happened; he said that some cattle had gotten lost in some brush and he'd gotten scratched up looking for them.

One day, to see why her husband kept coming home all scratched up, the woman quietly followed behind him. She saw him take off his clothes, walk into the lake, and emerge on the other side as a bull. She quietly took his clothes and carried them back to the house.

After that, for lack of clothes, the vaquero only went to the house at night, and did not enter into the house, but would stay out in the yard. One night, talking to his wife from outside the house, he told her, "Tomorrow they will take me to a bullfight, and I'll die. You take the *librillo* [an internal organ with plaits like a billfold] and you can live on the money it will have inside."

The next day, the woman followed the bull to the bullfight, weeping. After the bull was killed, she asked for the librillo, as he had told her to. There was a lot of money in it. (Fieldnotes VIII, 588–593)

Aurelio Condo, who told me this story, told me he had heard it from José Pillajo. When José Pillajo himself told me the story, he described the vaquero as captured or possessed by the devil, and he linked the wealth inside the librillo to the devil. Taita José's version also gave the *amos*, the owners (or renters) of the hacienda, a prominent role in the denouement. Someone—by implication, the vaquero's wife—informs them of the vaquero's self-transformation into a bull.

The amos decided, "Well then, we must take that bull to town, to have a bull slaughter." The bull would be killed in a bullfight in a fiesta.

... In anticipation of this, he [the vaquero/bull] told his wife, "... They'll kill me, but you come to get the librillo. Ask the amos for the librillo, telling them to give it to you as the vaquero's wife."

In Taita José's version as in Taita Aurelio's, the vaquero's wife is given the librillo and finds it full of riches.

Several points in the story are worth commenting on. Normally, an indigenous man would receive breakfast from his wife, and this gift of food symbolizes their relationship. In leaving the house without breakfast, the vaquero has already taken a step away from a domestic, human existence. Up in the mountain, he takes off his clothes, his "social skin" (Turner 1991), in order to undergo what might be seen as a sort of reverse baptism in the devilish lake. His wife then takes an important, ambivalent role in his fate: She takes away his clothes, thereby cutting him off more definitively from a normal, domestic existence—from this point on (in Taita Aurelio's version) he cannot enter the house. In Taita José's version, it is apparently the vaquero's wife who tells the amos of his daily transformation into a bull; yet she weeps as he is led away to be killed in the bullfight. The amos have decided his death, but under their noses, his Runa wife is able to secure wealth. This is a story of the social incorporation of wealth gained from the wild, from the devil, a process of social incorporation that comes about both through and at the expense of the vaquero, and in which the amos also play a role, though without being aware of the wealth in the bull's librillo. The vaquero is a strange sort of hero, one who gives himself over to the asocial wild, who pays for it with his life, yet whose last, perhaps redeeming, act is to ensure that his wife will have a comfortable life with the wealth that comes from his sacrifice.

Men form a very different sort of relationship with cattle in their normal work life. As plowmen, yoking and steering oxen, they emulate a different hero: somewhat surprisingly, Jesus Christ. This is surprising because among Ecuadorian mestizos, as in much of the Catholic world, men do not generally seem to view Jesus as a strong male role model. Armando Yépez told me this Quichua story, however, in which Jesus models and teaches male work skills.

Lord God... taught [people] to plant this earth, to plow, yoking the oxen.
...

[Some people]... are eating. He [Jesus] goes outside, and removing the ox's horn, enters dancing with it in his hand. The oxen are standing there, it's said, without horns, when it's time for them to be yoked.

So they are left awestruck, ... those people who were plowing. "Now what shall we do? This lord... takes off their horns," they say.

Mother Sta. María says, "Just be quiet. Now, he... mischievously takes off [the horns]. He himself will put back the horns in a little while." ...

So then, when it's time to plow, the oxen are all better. He yokes them, he plows. He plows a *cuadra* of land in just a little while, in almost no time. ... That's how he roams around, teaching. ...

It requires strength and mastery to control oxen when plowing. This image of Jesus as a skilled plowman, removing and replacing the oxen's horns as he teaches men to plow, seems to represent plowing as an act of masculine domination over the animals.[6]

Male work skills do emerge in oral accounts of the hacienda as a source of male pride and a feature of descriptions of exemplary men. In their work, again, men engaged with the wild. In this arena, though, rather than being wild themselves, they mastered and domesticated semi-wild animals—they taught horses to accept a saddle and rider, lassoed cattle, and subjected oxen to the discipline of plowing. Men sometimes recall proudly the praise they received from their hacienda overlords for such skills. Fiestas gave men an opportunity to display their skills and fearlessness publicly in a form of bullfighting whose object was to remove a blanket tied on the bull's back.

MATURE MASCULINITY: DOMESTICATION AND RESPETO

It was not only animal wildness that males confronted and mastered, however, but human wildness as well. Here again God provided a model. In baptism, as the baby was put in God's hands, it was transformed from an *auca,* a wild creature, into a *cristiano,* a human being—a transformation sometimes said to "cut its tail." The church wedding invoked God's help in creating or sanctifying a domestic union and maintaining tranquility between husband and wife. In both Catholic Penance and local forms of ritual discipline, God tamed and purified people through a human intermediary. In *pascuanchina,* a ritual whipping during Holy Week, it was said to be God who administered the whipping through an elder. Somewhat like a priest administering Penance, the elder admonished the person to be whipped and concluded with a blessing signifying forgiveness and purification.

As a man matured, married, took on fiesta sponsorships, and earned others' respect, he increasingly qualified himself to act as God's intermediary or aid in teaching and taming his juniors. By admonishing and ritually whipping them, he would instill respeto (respect), that is, a disposition to heed elders and follow moral norms. In connection with respeto, the elder/junior opposition was more salient in some ways than the male/female opposition. To judge from the language of oral historical accounts, men did not instill respect as *caricuna,* as males, but as *yuyaccuna,* "elders"— etymologically, "thinkers"—and in some contexts respected female elders (such as the wives of male religious authorities) also admonished and maybe punished their juniors. The ideal man in Pangor was a *yuyac,* a man "of respect," not simply a cari—a distinction somewhat similar to that between *hombre* and *macho* which Ana Alonso finds in northern Mexico (chapter 5, this volume). Male strength did nonetheless remain significant, perhaps particularly as a qualification for administering a ritual whipping.

Men could aspire to this mode of mature masculinity and to the position of authority within the hacienda community associated with it—that of the respected man who instilled respeto in his juniors, resolved conflicts among them, and helped to make them into responsible, domesticated social persons. There was nothing inherently contradictory between this form of indigenous male authority and indigenous subordination to the hacienda. To the contrary, by and large, the men who attained the highest positions of respect were pillars of the hacienda social order.

These positions were associated with religious functions. The *regidor* and *fundador* supervised important fiestas and fiesta sponsorship, mediated between the community and the parish priest, and helped mestizo hacienda authorities in judging domestic quarrels and other conflicts among hacienda residents. Hacienda landlords and stewards often named the *regidor* or *fundador* as overseer.

The authority of mature indigenous men could thus be incorporated as one link in a chain that extended up to the landlord himself. Landlords themselves, by extension, were like elders; the paternalism of colonial racial ideology was thus institutionalized in hacienda social structure and ideology. Reinaldo Sisa comments, "the *patrones,* the owners of the land—they were like our parents. Whatever they ordered, we did it."[7]

AUTHORITY ON THE HACIENDA IN
COMPARATIVE AND HISTORICAL PERSPECTIVE

In Ecuador, even before the 1895 Liberal Revolution, the relationship between landlords and laborers was theoretically based on a voluntary, contractual exchange known as *concertaje.* The landlord advanced money or goods to the laborer, who agreed to work for the landlord for a specified period or until the debt was paid off. Liberal regimes after 1895 introduced some modifications in local relations of power and authority, most significantly abolishing imprisonment for debt in 1918. Liberals also attempted to reduce the power of priests and to do away with religious offices such as that of *regidor* (see Guerrero 1991; O'Connor 1997).

In practice and in local understandings, however, power and authority on the hacienda were always based on much more than any voluntary agreement, and liberal ideology and legislation did little to alter this reality (compare to Alonso, chapter 5, this volume). Alongside the idea of voluntary contracts were other notions of authority with colonial and precolonial roots, notions that persisted to the end of the hacienda period. Hacienda landlords sometimes viewed themselves in terms similar to the Spanish conquerors—their mission was to subdue, civilize, and Christianize recalcitrant and brutish Indians (see Gose 1994). One prominent member of the Chimborazo landed elite summed up their tools this way: "the whip in one hand and in the other, bread, the gift, the cup of alcohol" (Borja 1953, 299). Like Runa hacienda residents climbing the ladder of respect, hacienda landlords were also capable of displaying generosity on festive occasions, adding to the sense of obligation that residents' use of hacienda land entailed. Landlords and stewards also joined with indigenous elders in judging conflicts among residents and imparting ritual discipline and moral instruction with religious overtones.

Together, these practices of authority constituted the subordination of hacienda residents to their bosses and of junior Runa to their seniors in distinctly unliberal ways. Subordination was not a choice on the part of rational, free subjects, but rather, the inescapable result of dependence, obligation, racial domination, and juniors' need for guidance and correction by their seniors. Moreover, in contrast to legal-rational forms of authority, authority based on debt or moral guidance was inextricably bound up with personal qualities, relationships, and histories of interaction.

The role of violence reflected these general characteristics of authority. Ritual discipline, in particular, embodied the personalized and nonvoluntary nature of authority on the hacienda. Parents might subject a rebellious son to a ritual whipping at the hands of a respected elder, for example, so as to impress dramatically upon the boy his proper position within the social hierarchy and the cosmological order. They also hoped that some of the elder's desirable personal qualities would thereby be transmitted almost physically to the youth (or "rub off" on him, as we might say) through the whipping. "This is respect," one informant recalled elders saying as they raised the whip.

While notions of respect could thus support certain forms of violence, however, respect also sometimes provided an ideological basis for contesting violence on the estate or within the household. Respeto implied hierarchy, but it also implied social harmony, mutual consideration, and a Christian moral order. When laborers today recall hacienda bosses' proclivity to violence, they sometimes speak of the bosses' lack of respeto as well as their own ability to cariyana. The rhetoric of respect also addressed domestic violence. Abused wives or their kin or neighbors sometimes brought complaints against husbands to Runa elders and mestizo bosses, who would inquire into the case and discipline the husband if they judged him to be at fault.

AUTHORITY AND RESISTANCE UNDER THE
HACIENDA AND TODAY

Thus far I have suggested that the term "cariyana" took much of its meaning from its semantic opposition to *respetana* (to respect) and *respeto* (respect). "Respect" referred to a hierarchy of authority, extending from parents and other senior kin up to indigenous religious authorities and mestizo hacienda bosses, built symbolically on the right of elders to tame their juniors through discipline. "Cariyana," by contrast, referred to the insubordination of "wild" juniors. The symbolic spaces of respeto were the chapel, where hacienda residents met once a week to pray and resolve quarrels; the hacienda courtyard, where elders administered discipline; and the home, where men and women were supposed to maintain harmony, where children first learned dependence and respect for elders, and where *pascuanchina* was also practiced. Cariyana was the invasion of wildness from the asocial spaces beyond human habitation into these realms and into human relationships. In moral instruction, the rhetoric of respeto was addressed to males and females—one of its prime objects was their complementary roles in a domestic union—while cariyana was masculine (even though both females and males could "get male"). If respeto was the ideological and experiential manifestation of a hegemonic alliance of hacienda bosses and indigenous elders, cariyana represented both a challenge to that alliance and an indication of its power to shape the meanings of such a challenge.

What were the social implications of this way of understanding resistance? On the one hand, to call open defiance of authority "cariyana" would seem to suggest that men could more or less be expected to engage in such defiance, at least at some stage in their lives. Certainly some men did occasionally challenge hacienda authority, as did some women, even though most informants say that fear made open defiance infrequent.[8]

On the other hand, the respetana/cariyana pair provided hacienda residents with a way to talk and think about authority and defiance that—together with violence and fear—tended to channel such talk and thought away from any notion of resistance as a potentially collective ethnic or class project. Respetana/cariyana constructed resistance as the spontaneous wildness of youths, especially males, who were not fully socialized. Even in retrospect today, while indigenous ethnic organizations promote a new narrative of "500 years of indigenous resistance," informants who tell of engaging in cariyana under the hacienda tend to reject any suggestion that they learned resistance from their elders. They stress instead their own willfulness, anger, masculinity, or other personal qualities. Their accounts also indicate that fellow laborers often criticized those who openly defied hacienda authority for their lack of "respect." Such criticism did not necessarily represent a sincere conviction, let alone one freely arrived at. Whatever the motivation, however, the point is that the language of respeto and cariyana easily reflected and reinforced disunity.

The moral ambiguity of cariyana and its resonance of wildness (even when justified) come ultimately out of a sense that authority is necessary for society. People today can speak of cariyana under the hacienda as a justified response to abuses and can take pride in their ability to cariyana, but it's hard to speak of cariyana as the basis for a long-term social project. Broadly speaking, indigenous hacienda residents lacked the autonomy to develop forms or discourses of authority that would present an alternative to the hierarchy of respeto and escape entanglement with the hacienda. When they weren't able to keep their own conflicts within private family circles, they were forced to recur to the authorities endorsed by the hacienda for resolving them.

Since the 1960s, autonomous villages have taken over the space formerly occupied by the hacienda, and they form the institutional base for the contemporary indigenous movement. But communal authority faces dissension from within and the competing jurisdiction of state-appointed civil authorities and the town-based legal system from without. The indigenous movement has responded with a discourse that links indigenous self-assertion to support for communal authority, thereby constructing resistance as a collective project growing out of local indigenous society.

Ironically, "respeto" is again a key term in this emerging discourse of ethnic revitalization and self-assertion, despite or even because of the ways "respeto" evokes the social order of the hacienda era. As always, "respeto" refers to authority and social harmony and to something taught and instilled by elders. But contemporary indigenous politics reconfigures respeto in significant ways. Now that indigenous villagers have gained communal autonomy, when they speak in village meetings of the need for more respect, they are looking for ways to strengthen the community's ability to maintain internal social order without recurring to mestizo authority. The contemporary discourse of respeto also supports efforts to build strong parish-level federations of indigenous communities and legitimates their demand that the mestizo state recognize their authority to deal with problems such as cattle rustling. Finally, a contemporary narrative of indigenous cultural resistance has given respect for elders a new meaning by casting elders as the ancestral bearers of authentic indigenous traditions. Respect for elders thus comes to mean valuing indigenous identity and culture, resisting the pressures some youth feel to shed indigenous ethnic markers, and joining in the collective project of building a political alternative to the structures and policies of the mestizo-dominated state.

Earlier I quoted an informant who speaks of indigenous laborers voicing demands before the state "like a male" (cari cari), but in the same breath he also refers to "knowledgeable" women as well as men gaining a public voice in the 1990s. The conception of the indigenous movement as a collective ethnic project creates some discursive space for women as well as men to participate and take leadership roles, while also shaping and limiting public discussion of gender issues in ways Cervone has discussed (chapter 8, this volume). The term "cariyana" does not appear a great deal in talk about contemporary political confrontations with the state. During the June 1990 National Indigenous Uprising, Pangoreño villagers applied the term *auca* to the soldiers patrolling the highway, constructing these representatives of the state as "wild," while they were demanding "respect."[9] The prominence of "respeto" exemplifies how contemporary indigenous political rhetoric in general constructs a collective, social project, a project of strengthening and changing society, not just resisting authority.

NOTES

The Social Science Research Council and the American Council of Learned Societies funded research in Ecuador from 1989 to 1992 through an International Doctoral Research Fellowship. The University of the South (Sewanee) supported subsequent research trips in the summers of 1995, 1996, and 1998. Writing was additionally supported by a Wayne State University Research and Inquiry grant in 2001. For helpful comments on the ideas and earlier versions of this chapter, I would like to thank David Frye, Mark Rogers, and the editors and reviewer of this volume. Thanks above all to the people of Pangor, especially the village that hosted my fieldwork, and to my parents and wife for their support.

1. See, e.g., Abu-Lughod 1990; Ortner 1995; Brown 1996.
2. Similarly, *huarmiyana* would mean "to get female," but I'm not aware of ever having heard this word in Pangor. Nor am I aware of any common or standard usage of *huarmishina* ("female-like") paralleling *carishina* (discussed below). The very lack or rarity of such usages might indicate that femininity as such does not carry so intensely ambivalent a symbolic charge as does masculinity.
3. See Allen and Garner 1997 for a similar observation on southern Peruvian Quechua speakers.
4. Other scholars have noted a broader association in the Andes between masculinity and speech (and between femininity and silence) in communal assemblies and other contexts. See Cervone in this volume, Harris (1980, 72–75), and several of the articles in Arnold (1997).
5. During the indigenous uprising of June 1990, however, I did hear some enthusiastic talk about fierce bulls in Salasaca whose indigenous owners were said to have directed them to charge the police.
6. Plowing is also understood as a masculine act in relation to the earth (*santo suelo*). It is one of the few activities that women are normally barred from by a taboo, a notion that the earth would take fright.
7. For a fuller discussion of the hacienda "respect" complex, see Lyons (n.d.); for an examination of "respect," authority, and punishments in relation to contemporary developments, see Lyons (2001).

8. Covert resistance was much more pervasive. See Lyons (1994, chapter 4) for an exploration of such resistance and its meanings.
9. Yet this was a fluid moment in which meanings and associations could easily shift. See note 5 for a contrasting reference to wildness in the service of the Uprising.

REFERENCES

Abu-Lughod, Lila. 1990. "The Romance of Resistance: Tracing Transformations of Power through Bedouin Women." *American Ethnologist* 17(1): 41–55.

Allen, Catherine J. 1988. *The Hold Life Has.* Washington: Smithsonian Institution Press.

Allen, Catherine and Nathan Garner. 1997. *Condor Qatay: Anthropology in Performance.* Prospect Heights: Waveland Press.

Arnold, Denise Y., ed. 1997. *Más allá del silencio: Las fronteras de género en los Andes.* Tomo I. La Paz: ILCA/CIASE.

Behar, Ruth. 1993. *Translated Woman: Crossing the Border with Esperanza's Story.* Boston: Beacon Press.

Borja, Luis Alberto. 1953. *Cabalgando sobre los Andes.* Buenos Aires: Ediciones Peuser.

Brown, Michael F. 1996. "On Resisting Resistance." *American Anthropologist* 98(4): 729–749.

Catta, Q. and P. Javier. 1994. *Gramática del quichua ecuatoriano.* Quito: Ediciones Abya-Yala.

Cusihuamán G., Antonio. 1976. *Gramática quechua: Cuzco-Collao.* Lima: Ministerio de Educación, Instituto de Estudios Peruanos.

Gose, Peter. 1994. "Embodied Violence: Racial Identity and the Semiotics of Property in Huaquirca, Antabamba (Apurímac)." In *Unruly Order: Violence, Power, and Cultural Identity in the High Provinces of Southern Peru,* edited by Deborah A. Poole. Boulder: Westview Press.

Guerrero, Andrés. 1991. *La semántica de la dominación: El concertaje de indios.* Quito: Ediciones Libri Mundi, Enrique Grosse-Luemern.

Haboud de Ortega, Marleen et al. 1982. *Caimi ñucanchic shimiyuc-panca.* Quito: Ministerio de Educación y Cultura, Pontificia Universidad Católica del Ecuador, ILL-CIEI.

Harris, Olivia. 1980. "The Power of Signs: Gender, Culture, and the Wild in the Bolivian Andes." In *Nature, Culture and Gender,* edited by Carol MacCormack and Marilyn Strathern. Cambridge: Cambridge University Press.

Harrison, Regina. 1989. *Signs, Songs, and Memory in the Andes: Translating Quechua Language and Culture.* Austin: University of Texas Press.

Lyons, Barry J. 1994. "In Search of 'Respect': Culture, Authority, and Coercion on an Ecuadorian Hacienda." Ph.D. diss., Department of Anthropology, University of Michigan at Ann Arbor.

———. 1999. "'Taita Chimborazo and Mama Tungurahua': A Quichua Song, a Fieldwork Story." *Anthropology and Humanism* 24(1): 1–14.

———. 2001. "Religion, Authority, and Identity: Intergenerational Politics, Ethnic Resurgence, and Respect in Chimborazo, Ecuador." *Latin American Research Review* 36(1): 7–48.

———. n.d. "Discipline and the Arts of Domination: Rituals of Respect in Chimborazo, Ecuador." Unpublished ms.

Múgica, P. Camilo. 1979. *Aprenda el quichua: gramática y vocabularios.* Quito/Pompeya-Napo: CICAME, Prefectura Apostólica De Aguarico.

O'Connor, Erin. 1997. "Dueling Patriarchies: Gender, Indians, and State Formation in the Ecuadorian Sierra, 1860–1925." Ph.D. diss., Department of History, Boston College.

Ortner, Sherry. 1995. "Resistance and the Problem of Ethnographic Refusal." *Comparative Studies in Society and History* 37(1): 173–193.

Thurner, Mark. 1993. "Peasant Politics and Andean Haciendas in the Transition to Capitalism: An Ethnographic History." *Latin American Research Review* 28(3): 41–82.

Turner, Terence. 1991. "Representing, Resisting, Rethinking: Historical Transformations of Kayapo Culture and Anthropological Consciousness." In *Colonial Situations: Essays on the Contextualization of Anthropological Knowledge,* edited by George Stocking. Madison: University of Wisconsin Press.

3. Women's Sexuality, Knowledge, and Agency in Rural Nicaragua ⤙

Rosario Montoya

WOMEN'S SEXUALITY, KNOWLEDGE, AND AGENCY

I heard the rumble of the jeep and looked up just as the vehicle swerved onto the dirt road that was the entrance to my hosts' house. A quick scan of the passenger seat and my eyes darted toward Luisa. She had already seen him. Our eyes met in a knowing glance. Slowly, as if having seen and felt nothing, she retreated into the kitchen. Knowing I couldn't do the same, I remained where I stood, awkwardly awaiting my host, Luisa's father Justino, who was grinning in my direction. As he climbed out of the car I smiled back as best as I could, but was unable to bring myself to as much as look at Juan, who had already climbed out of the car and was studiously ignoring Doña Miranda, Luisa's mother, and me.

For the third time that month Justino had brought Juan, Luisa's husband, from Managua, where he was training for bee-keeping, to spend the weekend with Luisa. Justino did this even though he knew full well that Luisa had told Juan repeatedly that their relationship was over. Later, Justino would tell me that he had done this for Luisa's sake, so that she "wouldn't be going from man to man." Luisa's relationship with Juan was, in fact, her first relationship. Justino's expectation that, if this relationship failed, his daughter would be going "from man to man" in all likelihood was a response to recent gossip about Luisa's love affair with Ricardo, a young man from a neighboring village. Indeed, from Justino's perspective, Luisa's affair must have only seemed to confirm the common expectation that young, unattached women would sooner or later involve themselves with a man; in Luisa's case, he suggested, Ricardo would be the second in a series of men.

Justino's statement invoked key aspects of gender ideologies in the southwestern Nicaraguan village of El Tule.[1] First, women who had been sexually involved with more than one man were supposed to be considered spoiled goods (*ya no sirven*), both in the sense of being immoral and of no longer of any use or value. These women were regarded as "bad" women or "women of the street," terms Tuleños used in contrast to "good" women—"women of the house" who followed proper gender and sexual comportment. In claiming that Luisa would be "going from man to man," Justino conjured up both meanings associated with "bad women": he questioned her sexual morality and presented the consequences that presumably followed the social evaluation of her loose morals, that is, it would be unlikely that she would ever again be approached by a man with serious intentions.

Luisa's unfolding story would prove Justino's predictions and his underlying assumptions about the fate of "bad" women to be wrong. In succeeding weeks, Luisa's relationship with her lover Ricardo intensified in tandem with gossip about the affair. Her reputation at stake, Luisa told me in no uncertain terms that time was running out: It was imperative that Ricardo "take her home" to live with him as soon as possible, thus formally making her his wife.[2] During this waiting period, she updated me periodically on Ricardo's efforts to "soften" his parents, who were long-time enemies of Justino's family and would not easily accept her in their home. But his delays worried her, and she occasionally—and bitterly—expressed doubts about the seriousness of his intentions: "I already told him that he can leave me if he wants, because in any case he already got what he wanted." Luisa's fears were unfounded, however. Ricardo was, like herself, in the throes of infatuation. Not long after our conversation he took her to live with him. Gossip about Luisa abated and then died down altogether, as she gradually reestablished her status as a "woman of the house."

Luisa's story is one of many stories of Tuleño women who were able to successfully maneuver their way into positions of respect in local society, even though they had failed to follow dominant prescriptions for appropriate female behavior. In this chapter, I draw on these stories to examine how Tuleño women exercised their agency from a position of social subordination. Specifically, I look at how these women, in the face of limited life options, used their sexuality—understood as a particular cultural and historical practice—to negotiate and even subvert social constraints on their sexual desire and choice of conjugal mate. While focusing on women's practices, my ethnography also makes clear that women's agency was necessarily constituted in relation to men's own agency as disciplining patriarchs and competing suitors who simultaneously opposed and colluded with women's transgressions.

According to dominant gender ideologies operating in El Tule, women who were labeled "of the house" or "of the street" were supposed to have *earned* these statuses through their behavior, particularly their sexual behavior. That is, a woman's place, understood as a more-or-less-permanent social status, resulted from her place as a set of appropriate or inappropriate practices. Most importantly, these statuses were supposed to be fixed once-and-for-all, for a woman's history was supposed to be inscribed in her body and could therefore never be undone. The fixity of these ideal trajectories thus reflected the putative binary morality of Tuleño womanhood.

Tuleño women's stories, however, reveal that such fixed life patterns were rarely borne out in practice. Contrary to what Tuleños claimed, women regarded as "bad" were often able to become "women of the house"; women regarded as "good" women of the house rarely had irreproachable sexual pasts; and the dubious sexual practices of "good" women in the present did not, in themselves, necessarily or forever earn them the designation or presumed fate of "bad" women. Moreover, these stories demonstrate that women were not only knowledgeable about this state of affairs but also that they played a key role in its construction.

By means of their transgressive sexuality, Tuleño women simultaneously legitimized and contested the basis for their proper place in local society. They legitimized the status of domesticity by seeking to eventually become "good" women of the house; but they also contested the basis for this status by challenging the parameters within which women could legitimately exercise their conjugal choices and act on their sexual desires and the ways these choices were to affect their future conjugal possibilities. That is, they contested the ways in which women's place was to be linked to their sexual behavior. In this chapter I show how Tuleño women, by partially penetrating (Willis 1977, 119–125) the workings of local gender ideologies, crafted the conditions of possibility for their own mobility across social statuses that were supposed to be impermeable to one another.

My interest in women's agency in the space of the home emerged in the course of ethnographic research as I found that Tuleño women did not fit the models of female agency that prevail in feminist research on Latin American mestizo women. While this research has produced rich accounts of women as political actors and workers,[3] its concern with women's experiences in the home has, for the most part, been limited to how women's "traditional" identities, especially as mothers, bear on their engagement with the public domain. Thus, heterosexual women's so-called domestic practices, including their conjugal relations and sexual practices, have been relegated to the margins of feminist analyses of Latin American women's agency and transformations in their gender consciousness.[4] Yet while the public/private (or public/domestic) distinction certainly exists as an ideological feature of official cartographies, women do not necessarily live their lives according to this distinction (see Stephen 1997); nor is their gender identity reducible to the terms it establishes (see e.g., Diaz-Barriga 1998, 2000). Rather, because women move across various realms mapped by both local and official cartographies, the significance of their practices is constituted in relation to any number of realms which may exceed, or not be captured by, the public/private distinction. Thus a fuller understanding of women's conjugal and sexual experiences is called for—both for what it can tell us about the potential of these experiences to bring about gender transformation, and to enhance our understanding of how women *do* engage with arenas such as politics and waged work.

In this chapter, I demonstrate how women can be agents, and perhaps even agents of change, in and through their conjugal and sexual practices. I understand agency to be the capacity to respond to situations of misidentification or contradiction in creative and unanticipated ways (McNay 2000), and with a sense of authorization (Ortner 1996, 17). To the extent that subjectivity involves a set of embodied potentialities which are lived through under concrete material conditions, agency is always emergent in a determinate historical praxis.[5] From this perspective, I examine the sexual practices of women who broke with prevailing gender norms to carve a space for greater conjugal choice and sexual desire. While I am mindful of the view that norms and transgressions are mutually defined such that the latter are always implicated in cultural systems of power (Foucault 1978), I argue for a more nuanced, dialectical understanding of the relationship between norms and practices that focuses on the *tensions* between them and looks to the possibilities for transformation that emerge therein, as these play out within specific formations of power.

My goal is to expand the terrain of research by tearing down—*desalambrando*—conventions guiding much feminist scholarship on Latin America that locates mestizo women's agency, power, and/or resistance almost exclusively in the public domain. Thus, rather than assuming the (in)significance of women's conjugal and sexual practices, I examine their logics from women's own points of view as subordinate actors knowledgeable about their conditions of existence. This analytic stance allows me bring to light the creative dimensions of women's transgressions and their effects on gender/sexual relations, both of which would be obscured by the imposition of the analytic categories of public/private on Tuleño women's practices.[6]

SPACE, GENDER, AND THE SEXUAL/MORAL ECONOMY IN EL TULE

El Tule is an agricultural and cattle-raising community of about 400 people spread over ten square kilometers in northern Rivas, some five kilometers inland from the Panamerican highway. Community lands include private family parcels in principle inherited in equal shares by sons and daughters, and cooperative land received during the Sandinista revolution (1979 to 1990). The settlement pattern is scattered, although extended families tend to cluster on their own lands. The distance between village homes or home clusters rarely exceeds five to ten minutes by foot.

Under the Sandinista and neoliberal governments of the 1990s, as in the past, most Tuleños' daily lives transpired within the bounds of their village and neighboring villages. This does not mean that El Tule was or had ever been isolated: Particularly during the Sandinista decade, it was tightly integrated into the revolutionary process, earning a place as a "model" or showcase Sandinista village visited by hundreds of revolutionary tourists.

Tuleño men were and had always been unencumbered by social limits on their movements across space, regardless of time of day or night. Not so with women, who even in the 1990s continued to be subject to social restrictions on their movements. On a daily basis most women left their houses only to go to the well, situated a close distance from their homes, and only if their daughters were too young to haul water themselves. The few exceptions included two single older women who made a living in part by selling goods obtained from nearby market towns; one or two women who periodically visited daughters living in this town; and, since the mid-1990s, a handful of young women who traveled to the municipal capital to attend secondary school. The restriction of women's labor to their homes was tied to ideas about patriarchal prerogatives and obligations: Only in times of male unemployment should women look for wage work outside the home, usually as domestic servants in nearby towns or cities.

The changes wrought by the Sandinista revolution—which included the building of a school, a health center, and a road running through the village and connecting it to the Panamerican highway—expanded women's range of movement to include occasional visits to these facilities and, in emergencies, to the hospital in the provincial capital. The construction of the road, in particular, brought opportunities for greater mobility on a regular basis, as the new ease of travel in full view of the community enabled women to run errands to other homes without spousal conflict. Membership in the Sandinista women's organization *Asociación de Mujeres Nicaragüenses Luisa Amanda Espinoza* (AMNLAE) also increased women's mobility outside the home through participation in mobilizations and women's collectives (see Montoya forthcoming). Finally, with electrification in 1989, entertainment opportunities, which in the past were restricted to occasional celebrations in people's homes, had also expanded—at least for unattached women—to regular nightly gatherings in homes with televisions. Despite these avenues for increased mobility, women knew that, to protect their reputations, they must leave their houses sparingly and with proper justification.

The confinement of women's movements to sanctioned times and places can be seen as a kind of "ideological work of gender" (Poovey 1988, 2–3) that created and sustained relations of male domination. This work was ideological in that it entailed mobilizing gendered representations and practices in ways that naturalized asymmetrical relations of power between men and women. A key medium for carrying out this ideological work was the language Tuleños used to delineate the statuses people could occupy in village society, to identify practices appropriate to each status, and to locate these practices in particular spaces. These norms were reproduced through being somatized and practiced in villagers' daily lives, even as particular contradictory dynamics rendered this reproduction always uneven.

For women, the two available statuses established by the dominant ideology were those of "good" and "bad" women, terms that corresponded to the spatialized designations of "woman of the house" and "woman of the street." These terms differentiated women primarily by how they exercised their sexuality and evaluated them in both normative and moral terms. Tuleños posited the domains of house and street as discrete, mutually exclusive, and highly gendered. These ideological constructions of place shaped women's goals by subordinating them to men under the terms of a specific sexual-moral economy between husband and wife. Becoming a "woman of the house" entailed a public recognition that she was now a woman with a man's backing—a protected woman—and therefore one who must be respected. As Tuleños would say, "El hombre hace valer a la mujer" (men give women value). Ideally, this position conferred economic stability as well as status. Women's roles in the house, then, were cast within the terms of a classic patriarchal "bargain" (Kandiyoti 1991): In return for protection, respect, and economic stability, the wife was to uphold the respectability of the house by keeping to a set of clearly prescribed practices. Most importantly, Tuleños would say, she should work hard and serve her husband graciously; bear children, using contraceptives only with his permission; stay out of his doings outside the home; and be sexually available and faithful to him, as men's honor hinged on their wives' sexual fidelity.

While Tuleños defined the status of woman of the house by a set of activities appropriate to that domain, these activities acquired their meaning in contrast to those activities associated with the domain of "the street"—a highly sexualized place where only men and "bad women"— women unsuitable for a respectable social position—were supposed to venture. Like "the house," "the street" was more than a geographic location; it was a culturally constructed concept that stood for all practices related to sexual conquest. Hence husbands were careful to protect their honor via their wives' restriction to "the house." From the perspective of women of the house, then, the street functioned as a disciplinary mechanism for keeping them in their "place," in which "place" refers both to a particular physical location and set of appropriate practices.

The ideological distinction between different kinds of women, which rested on the casting of house and street as separate and separable spheres, also effected a separation within men's activities such that they could legitimately occupy both house and street. That is, a man had obligations in the house as responsible patriarch *and* was free to pursue sexual (and related) activities in the street as philanderer. For good women, this translated into an injunction to be indifferent to, or at least ignore, men's affairs in the street, for these were supposed to represent no threat to wives.

Most of the time women (and men) invoked dominant sexual/moral norms to assess their own and other people's actions. This did not mean, however, that everyone at all times believed in these norms unambiguously or that people adhered to them in practice. Rather, it meant that these norms fostered embodied investments by providing the only publicly accepted moral standards and criteria for conferring gender value. They thus structured both the constraints under which women (and men) operated and their motivations for seeking prized statuses during most of their lives. That these norms were dominant in El Tule did not mean that villagers were ignorant of other ways of organizing sexual life. On the contrary, Tuleños had been exposed to a variety of sexual cultures through contact with urban and middle-class people during and after the revolution and through their exposure to the radio and TV. Nonetheless, well into the 1990s, they generally[7] continued to invoke and apply local sexual norms as though these were the only conceivable option. This had less to do with small-town parochialism than with men's constant policing of the mechanisms by which they retained their dominance and control over women (Montoya forthcoming).

WOMEN'S SEXUALITY AND "THE HOUSE" UNDER CHANGING HISTORICAL CONDITIONS

Every Tuleño woman I knew aspired to the status of "woman of the house." This aspiration was consonant with the terms by which women were accorded prestige and value within the conjugal sexual/moral economy discussed above. Nonetheless, during the 1980s and 1990s, various processes came together to create conditions for a critical mass of Tuleño women to engage in illicit sexual activity. As a result, throughout these decades the village's sexual/moral economy operated primarily through the transgression and not the norm.

My historical research suggests Tuleños' sexual/gender arrangements had always given rise to contradictions in spousal relations. Since at least the 1970s, and especially in the 1980s and 1990s, these contradictions were, in part, tied to the partial erosion of the economic basis for these arrangements, which made it difficult for women and men to comply with their part of the patriarchal bargain. Several factors contributed to the decline of the regional agrarian economy: land pressures, increased reliance on purchased items, and, under the post-Sandinista neoliberal regimes, shortage of agricultural credit. These dire economic conditions rendered men less able to properly support a family, much less cover the economic exigencies of a second woman. For many women, men's diminished economic capacity and the economic and social threat posed by men's female lovers[8] were a source of great dissatisfaction with their marital situation. While most women whose conjugal unions lasted more than ten years were able to attain conjugal stability, some continued to perceive abandonment by their husbands as a real possibility well into their late thirties.

Women's sexual transgressions were part of a broader set of tactics through which women dealt with the contradictions generated by the disjuncture between their marital expectations and the actual economic strains and instability of their conjugal relations (Montoya 2001b). This is not to say that these conditions *caused* women's transgressions, but rather than they led to situations conducive to such transgressions. Indeed, my ethnography shows that women's sexuality had its own dynamics: Rather than satisfying strictly instrumental objectives, it was a source of ludic and erotic pleasure. Thus I suggest that other factors that brought national and global trends into El Tule may have contributed more directly to the gender/sexual landscape that I encountered during my fieldwork.

The first was the increasing sexualization and commodification of women's bodies that came with increased exposure to the mass media. A shift in women's fashions traces some of these changes: from simple, long, homemade cotton dresses worn until the 1950s and 1960s to purchased clothing increasingly attuned to contemporary international fashion. A second factor was the change in gender ideologies that began with the Sandinista revolution. Tuleño women were particularly influenced by initiatives and gender discourses espoused by feminists from various national organizations that came to work in the village. These initiatives promoted women's right to work outside the home, live free of domestic violence, refuse sexual intercourse with their husbands, and participate in decisions concerning birth control. While these feminists did not discuss, much less encourage, greater sexual freedom for women, it seems reasonable to surmise that their advocacy work expanded the possibilities for Tuleño women to reconfigure their relationship to their own body and sexuality in ways that contradicted dominant sexual norms.

Finally, women's illicit sexual practices were facilitated by various social conditions. First, since a large proportion of unions in El Tule were either endogamous or involved inhabitants of neighboring villages, most women could easily return to their parents' house if their unions failed.[9] Women's survival was thus not usually at risk were they to lose their partners. Second, and of key importance, the increase of "outside" spaces that women could legitimately occupy combined with population growth produced regular opportunities for women's unchaperoned interactions with men of marriageable status.[10] Finally, despite the extent to which people in general—and husbands in particular—obsessed over controlling women's movements, men rarely left women on account of their transgressions. How this is to be reconciled with the construction of masculinity as intolerant of women's infidelity is a point I have addressed elsewhere (Montoya n.d.). Here, suffice it to mention two factors that worked at cross-purposes with men's traditional masculinity: First, the circulation of a large number of women engaged in infidelity meant that despite what everyone affirmed, men did become infatuated with, and formalized relationships with, women who were not "good," and women knew this and used this knowledge to their advantage; second, traditional norms were not buttressed by severe enough sanctions for men (or women) to compel men to marginalize "bad" women who failed to uphold men's honor. Thus, although men's standing as patriarchs was measured in part by their ability to protect and demand the honor of their daughters and wives, if "their" women engaged in illicit sexual activities Tuleño men could save face by feigning ignorance. If the affair was too public, men found other ways to try to salvage their own honor. For example, Victor did little about his wife Carmen's extraconjugal relationship because he could not control her actions and he did not want to lose her. However, he once beat her publicly to assuage rumors about a presumed affair between Carmen and his niece's boyfriend, even though he knew the rumors were untrue. The important point here is that, while for men (as for women) reputation and honor did matter, there were ways to protect them in the face of women's sexual misconduct. This meant that, wittingly and unwittingly, men became enablers of women's transgressions.

REMAINING "GOOD": VISIBILITY, TIMING, AND THE MANAGING OF GOSSIP

As in other Latin American societies historically, for women in El Tule, maintaining a good reputation depended primarily on establishing one's sexual propriety publicly (rather than on actual deeds). However, women in courting or stable conjugal relationships faced obstacles that made it difficult for them to remain within the terms dictated by "house" norms. Thus women had to find ways of protecting their reputation while engaging in practices regarded as morally questionable to achieve their goals. In large measure, this was a question of skillfully gauging the moral risks women could take given the degree of "visibility" of their practices.

The notion of visibility referred to the degree to which people "saw" or noticed disreputable practices. Thus Tuleños would often comment that when women were sexually unfaithful, "es mas visto" (it was more noticeable and it looked worse) than when men were unfaithful. But while people generally spoke in absolute terms about the visibility of women's dubious sexual practices, in fact, the extent to which women's morally transgressive practices were visible varied depending on the woman's accorded value. In a sense, Tuleño women were assigned a "value fund" that figured centrally in the community's willingness to either "see" or look past the women's sexual transgressions. This value fund, based as it was on women's perceived sexual morality, could be sustained, eroded, or depleted, by the public assessment of women's sexual comportment.

As the stories below show, women's management of gossip through the skillful timing of both transgressions and conformity to dominant norms was key to intelligently husbanding their value fund. In my analysis of these stories, I stress the anticipatory dimensions of women's choices in order to shed light on the constitutive role of imagination and creativity in their agentive practices (see Ricoeur 1994, 126).

The case of Lorena illustrates the advantages of having a "full" value fund in El Tule's sexual/moral economy. Lorena was 14 when she and Emilio began courting. A few months later Enrique, Lorena's cousin and Emilio's good friend, confided in me that Emilio was in love with Lorena and wanted to "take her home" with him, and that he had told her as much. But Lorena, who had never been sexually involved with anyone, had decided to have her first sexual experience with Emilio before he actually proved his serious intentions by formalizing their relationship. Enrique told me of his conversation with Emilio once he learned Emilio and Lorena had been together intimately. "I told Emilio that now that he had made her a woman he could not leave her like this. He would be paying her very badly if he left her like this."

In Enrique's eyes, Lorena was worthy of Emilio's consideration because she was a woman with a "clean" sexual past. Her loss of virginity, in this case, was not too "visible" and could be morally forgiven not only because she had lost it for love but, most crucially, because Emilio was her first man.

Contrary to dominant ideologies, then, a woman could keep her value despite losing her virginity if her lover "took her under his charge." This suggests that it was not virginity itself that was of value, but rather the process of remaining a virgin until achieving—or, in this case, being in the process of achieving—the status of wife and mother.[11] This explains the urgency with which Enrique encouraged Emilio to formalize his relationship with Lorena.

Shortly after Enrique told me about these events, Emilio and Lorena got together formally. Timing was all-important here. By formalizing their relationship sooner rather than later, Emilio and Lorena preempted malicious gossip about her that would have begun to taint her reputation, eroding her value fund and reflecting badly on both of them. Lorena's full value fund, then, served her well in at least two ways: It facilitated Emilio's ability to take her home while avoiding community ridicule and it sheltered her transgressive practices from undue visibility long enough for her and Emilio to resolve their situation. Thus, through her efforts and Emilio's collusion, Lorena remained "good" in the community's eyes.

Women with courting stories like Lorena's, but whose first lover did not "take them home," found they had begun to erode their value fund. The main problem facing women from the start was men's practice of developing relationships with several women while having serious intentions with only one, if any. Thus a woman had to be acute in gauging her suitor's intentions. Men's words were not much help, for, as I was told many times, their declarations of love were often empty and merely aimed at "getting what they wanted." It was equally difficult for women to interpret men's simultaneous relationships since suitors tended to deny them or play down their significance if they were found out. Perhaps men's duplicitous solicitations encouraged the common practice among young unmarried women to develop flirtatious, sometimes even courting, relationships with more than one man. Indeed, this practice had shaped courting etiquette such that young women in the village rarely flatly rejected a possible suitor's offers. Instead, they were supposed to make up excuses for not being able to comply with their suitor's wishes while being careful to keep the door always open.

Playing the game of keeping options open,[12] particularly if these had developed into courting relationships, required skill and knowledge if a woman was to maintain her reputation. By the time Maira, another of Enrique's cousins, was 16, she had had several boyfriends, and it was rumored that she had been intimately involved with at least two of them. When she eloped with Elvin, I asked Enrique to tell me what he knew about the relationship since I had been under the impression that she was courting Humberto. "She was courting both of them at once and the other day she went to the [saint's celebration] in San Andrés and just dumped Humberto for Elvin," he explained. "What happens is that these days women no longer love the same way men do." Clearly oblivious to the fact that women could not depend on romance alone, he proceeded to call Elvin a fool because, he said, "that woman is no longer of any value."

Enrique's negative comments about Maira's value were consistent with that of other villagers, especially men, who routinely demeaned women engaging in sexually transgressive behavior. These comments worked as a disciplinary technology, shaming some women into pushing for a resolution of that particular relationship, or ending it. In this respect, Enrique's assessment of Maira was not unlike Justino's assessment of Luisa that I described in the introduction. Men's slanderous gossip, then, served as a prime medium for creating a more-or-less consistent public assessment of women's value *at that particular point in time*. What is important, however, is that women's value in El Tule's sexual/moral economy was not temporally fixed, since it did not depend on people's evaluations of sexual practices in the abstract, and even less on their evaluations of women's transgressions at the time in which the transgressions occurred. Rather it depended on whether or not these women were able, at any future moment, to be involved in a formal relationship. This is consistent with my observation that women of the house were more prestigious and were treated with more respect than women with comparable sexual histories who were single. It is also consistent with Tuleños' dictum that "men give women value." But of course there was a link, at least in principle, between women's reputations and their ability to achieve the status of the house. Thus women's tactics aimed not only to engage men's desire and love (through a combination of sexual enticement and by signaling "goodness"), but also to facilitate men's ability to formalize the relationship by protecting their own image and reputation through the management of gossip.

Contrary to Enrique's statement, Maira's case demonstrates the "play" that actually existed in the application of sexual/moral norms, and which women clearly counted on. Most importantly, there was no way of "proving" Maira's alleged past sexual relationships and this was essential to Elvin's ability to save face for making her his wife. But Maira had been pushing her luck as she courted two men at once, for I noticed that the more men a woman was rumored to have been with, the more likely people were to declare gossip "true" versus malicious or unproven talk. Judging by Enrique's comments and Elvin's decision to make her his wife, Maira's past had eroded her moral value fund to the point at which she could not always command charitable interpretations, but not enough to seriously diminish her options. Moreover, whether it was done at her insistence or of Elvin's own volition, the formalization of their relationship before yet another round of comments could begin about her degree of her intimacy with Elvin or Humberto was important. For given Maira's limited value fund, managing gossip was essential to her conjugal possibilities in a way it might not be in the life of a woman like Lorena. In any case, as with Luisa, once Maira's place as a woman of the house was established, people seemed to find her past less worthy of comment than they did when the outcome of her relationships was uncertain.

The reputation of women in conjugal unions was based on how good and sexually faithful they were as wives. These women operated under different constraints, primarily capitalizing on the prestige they already possessed by virtue of being women of the house. Like unattached women, young women in the early years of formal unions were often uncertain about the future of their relationships. In particular, they reported fearing being abandoned, mistreated, or neglected by their husbands. Women also commonly claimed to be unhappy about their husbands' poverty. Perhaps there was a link between these situations and the tendency among young women in conjugal unions to flirt with men other than their husbands. While women often seemed to carry on these flirtations simply as pleasurable games by which they continually tested their sexual appeal, they sometimes received gifts, the source of which they went to great lengths to keep secret. In other cases, these relationships developed into extraconjugal affairs and/or new marriages. (It must be said, however, that all the women I knew who were in love with their husbands remained sexually faithful despite how badly their husbands treated them or how bad their economic situation might have been.) Luisa's story, which I began to discuss in the introduction, reveals some of the ways women in conjugal relationships attained or retained a respectable position despite obvious transgressions.

Luisa claimed that she fell out of love with her first husband Juan because she was disappointed in his lack of economic motivation, exemplified by the fact that, several years into their marriage, he still had not built her a house of her own. While they were living at her parents' house, Luisa flirted with several men from whom she occasionally received small gifts. Then she fell in love with Ricardo, a young man who, ironically, was still economically dependent on his father.

Meetings between lovers in El Tule were practically impossible to keep secret for long. As gossip mounted, Luisa was under pressure to either give up the relationship or push Ricardo to "take her home," which she did continuously. Twice during this period, Justino warned her that people were "talking." "You need to prove that [the gossip about your affair] is a lie," he told her, anxious to keep her away from any situation that could generate further "talk." This was a moment of danger for Luisa, one in which gossip was crossing the boundary from what people perceived as malicious talk to what they believed was true. The resolution of the situation came from Justino as he attempted to protect his honor. After a party at the schoolhouse, he confronted Luisa: "I saw the way you were dancing with that man—that is proof enough for me that you are his woman! And the whole community saw you! Get out of this house right now!"

Luisa went to stay at her aunt's house in a nearby town. The next day Doña Miranda, Luisa's mother, told me that if Luisa knew what was good for her, she would stay away from Ricardo. This would "prove" to people that the rumors were false and eventually Justino would take her back. But things would not work out that way. Two days later, word reached our house that Luisa had been seen in Ricardo's house. She was formally his wife now. Luisa later told me what had transpired as all in her house waited, ears perked, for news of her next move. She had sent Ricardo a message warning that he either take her in now or that it was all over and he would have really messed up her life. As she expected—or hoped—he came to get her immediately.

Luisa's story illustrates several points about Tuleño women's tactics. Unlike many lover relationships in the village, which ended quickly under the pressure of community gossip, Luisa continued to see Ricardo, gambling on his sincerity. Once her father threw her out, however, she knew that she could wait no longer: She gave him an ultimatum, refusing to see him unless he formalized their relationship. Luisa's gamble paid off. However, things might have turned out differently, because Luisa, like other Tuleño women, was playing futures and was never certain how things would work out in the end. In this process, timing was of the essence, especially in relation to gossip: Luisa knew she must not let gossip mount too much or it would work against her value fund. This meant gauging at what point she had to leave the relationship or to what degree she could "push" Ricardo for a resolution.

The decisions Luisa made at various points to continue gambling on Ricardo also shows that she was aware that she could neutralize gossip's condemning effects, if he came through for her. As her own parents knew, because Luisa had a full value fund until she got involved with Ricardo, people would be willing to publicly label the talk about her affair an unconfirmed rumor if she behaved appropriately and stopped seeing Ricardo. Indeed, even after Justino threw her out, she still could have "proven" the rumor untrue by avoiding Ricardo from then on. But that outcome would have been less than desirable, for various reasons: First, Luisa loved Ricardo, and wanted to be his wife; second, her reputation would not have gone unscathed, especially once her father evicted her—a highly unusual step for a father (or even a husband) that only seemed to confirm the rumors; and finally, she would have had to suffer the unbearable embarrassment of having "been made a fool of" by Ricardo.

Luisa's story confirms my observation that the status in the house was a source of considerable prestige and moral capital, so that as a long as a woman moved directly from one position in the house to another, her value fund suffered little. In Luisa's case, as in others I witnessed, there was no way to hide her infidelity precisely because of having gone directly from one man to another. I was therefore surprised at the greater respect and social acceptance she commanded as Ricardo's wife than women whose sexual deeds differed little from hers, but whose gambles on their lovers had not paid off. Luisa's story is also important in demonstrating that desire bore centrally on sexual and conjugal decisions that may appear to be instrumentally aimed at leaving undesirable relationships while attaining (or retaining) the status in the house. (Indeed, desire/love was *the* key motive for every woman in these situations whom I interviewed.) For while Luisa entertained various options in light of her husband's perceived failings, she fell in love and left him for a man who could not offer her much more economic security than her husband. (In many cases, however, women's sexual transgressions were not conditioned by dissatisfaction with their spousal relations, but rather by desire/love for another man. This is particularly clear in the frequent cases in which women with exemplary husbands took lovers, and even left good mates for less sexually faithful or economically stable men.)

Still, in this case, as in others I witnessed, sexual desire and status concerns (i.e., becoming a woman of the house) were inseparable motives behind women's sexual practices. That is, although women who left their husbands usually did so because they fell in love with someone else, I don't know of a single case in which women did so if their lover did not offer to make them a woman of the house. Thus, desire, as a prime motivation for overrunning the fences of local sexual/moral norms, was at the same time subject to key constraints erected by these fences.

The stories in this section offer a glimpse into a more generalized aspect of gender relations in El Tule that belies dominant representations of "good" women who presumably observed sexual/moral norms: Namely, that in one way or another, the breaking of these norms was "normal" in El Tule. That is, even as women played the game of "remaining good" publicly, the transgressive practices of bad women were an intrinsic part of the configuration of "the normal"—a place where the dramatic scale of women's transgressions destabilized the boundaries between good and bad women.[13]

Importantly, the normalization of women's transgressions resulted from more than individual women's sexual misconduct. Men, as we have seen, directly colluded with women's morally dubious deeds even as they criticized other men when *they* fell in love with transgressive women. More broadly, young women in El Tule deliberately shaped this state of affairs as they commonly helped and encouraged other women to transgress, in some cases, only to reveal their transgression later. In one such case, Flor, a woman in a formal conjugal union who had a series of lovers, was trying to set Luisa up with a man while she was with Ricardo. When I asked Doña Miranda why she was doing this Doña Miranda astutely replied: "Maybe so that people can't say that only she does it [has lovers]." In cases such as this, gossip functioned not as a lubricant for custom or tradition (Lancaster 1992, 91) but rather as a medium for normalizing and thus enabling women's transgressions.

BAD WOMEN ON THEIR WAY TO THE HOUSE

I have been arguing that an important tactic through which women achieved the status of woman of the house was by timing their sexual practices so as to remain publicly good. Women whose status was that of woman of the street, however, did not have that option. Nonetheless, there were ways in which these women could render the boundary that separated them from good women more fluid and porous than what dominant ideologies implied. First, while women of the street were always short on prestige and value by virtue of occupying a position "on the outside," in some cases they could earn people's respect if they observed normative dictates concerning proper female comportment. Second, they could earn value if their lover demonstrated a serious investment in the relationship by providing economic support and spending time with the woman in and outside of her house.

But even women of the street whose lovers had no serious commitment to the relationship could counter their negative image in small ways, as Luisa's advice to her cousin Teresa suggests. Teresa's lover, Pedro, lived with his wife and children in a neighboring village. When Pedro visited his parents in El Tule, Teresa would spend the night with him, coming to his parents' house late in the evening and leaving early the following morning. Luisa recounted her advice to Teresa: "I told her that she shouldn't be so stupid, that it looks very bad that she goes [to his parent's house] just to sleep with him. And that she should at least stay in the morning and do the dishes so that she can show her propriety [*su formalidad*]." By performing the domestic practices of women of the house, that is, Teresa could demonstrate to her lover's family her commitment to this relationship, and thus improve their assessment of her.

Mayela's case illustrates how far women of the street would go to get "a piece of the house," so to speak, and thus achieve a measure of social value. For approximately two years, Mayela had a relationship with Denis, a man who lived with his wife Pilar and their children in his parents' house. During this time Mayela bore Denis a child. But although Denis continued to visit Mayela after their child was born, he never helped her economically or gave any other indication of formality in their relationship. Mayela, however, was determined to publicly establish her place in Denis's life. At the baptism party of Denis and Pilar's child, Mayela not only showed up but purposely danced provocatively with Denis, spurring Pilar to physically attack her. To my surprise, Denis's sisters intervened against Pilar, and Denis did not defend her. (Denis's behavior was not uncommon: Tuleño men would often set "their" women up to fight each other.) From then on, every time there was a party in Denis's parents' house, Pilar spared herself this situation by going to her parents' house until the following morning. Eventually though, she had to stay at the parties, having learned that, at the previous party, Mayela had spent the night with Denis in her own bed. While practices such as these certainly did not earn Mayela respect, she *was* able to publicly establish Denis's regard for her by competing somewhat successfully with Pilar for his attention. This earned Mayela a measure of prestige, if not value.

Most women in the status of the street, particularly those who were in stable relationships with men who were already someone else's husband, sought different ways to establish their value. These women generally considered themselves the man's (second) wife, bearing him children and behaving like women of the house. Depending on the formality of the relationship, the community sometimes considered these women quasi-wives or concubines. One such woman was Clara, the mother of four children with Roberto, a man in a formal union with Linda. According to Doña Miranda, Clara had been regarded as a good and hardworking woman throughout her relationship with Roberto, even though she was lacking in social prestige. Still, due to her perceived sexual fidelity to Roberto and the relative stability of their relationship, people referred to Clara (and other women in positions like hers) as the woman Roberto had "on the outside" (la mujer que tiene por fuera), an expression denoting a relationship somewhat more formal than those referred to as "of the street."

While most women, and especially women of the street, strived to obtain social value, the failure to do so did not necessarily mean that women could not attain the status of the house. In these cases, the importance of women's "goodness" and/or their arts of seduction is clear. Clara's story is a good example. After years of behaving as a proper wife should without receiving much from Roberto, Clara gave up playing the "good woman." While I lived in the village, she and her sister (who was in a similar situation) were among only four local women who openly drank beer at the village bar, and I heard from many people that she routinely slept with men who were obviously not making any commitment to her. Despite her sullied reputation, when I returned to the village two and one-half years later, I found out that Clara and Chema, a very good and respected man whose wife had recently left him for another man, had become a couple and had had their first child.

While women of the street routinely enticed single men to formalize their relationship, they did so with men in formal unions as well, although rarely with the same success. Maribel was a very attractive woman from a nearby village who had separated from her husband and begun to live what villagers considered to be a dissolute life. She was infamous for frequenting bars, where she would often become drunk and rowdy. It was said that she had many lovers during this time. In one of her outings, Maribel met Francisco, a man from El Tule who was widely feared for being brave, unusually aggressive, and "very manly" (muy hombre). Once she got involved with Francisco, she waged war on his wife Guillermina. As was common in such cases, Francisco attempted to keep both women. "He would come here and stay with me sometimes," Guillermina explained, "but that woman would come looking for him and he would leave with her." Eventually, Maribel forced Francisco to make a decision, which he did in her favor, leaving behind Guillermina and their five children.

These stories reveal facts about Tuleño women and their relationship to the street that contradict the dominant ideology: Far from being doomed, women of the street could become "good women." In fact, most women of the street did become good women at some point. Far from being a space of male dominance in which men "conquered" unwitting victims or played with "bad women," the street was a place in which women too "conquered" their men and thus enhanced their social position. Nor was the street a space closed off from the respectable space of the house, one that only males and bad women inhabited. Rather, it was a space continuous with the house, inhabited by women potentially on their way to the house. The fluidity of borders between house and street meant that the definition of these places, and of good and bad women, were always unstable and contested (Massey 1994, 2–5).

What happened to these "bad" women once they were in the house? The answer to this question reveals with crystalline clarity the way in which the dominant ideology worked in El Tule. If, once in the house, a woman behaved properly, her status was so prestigious as to override her "questionable" sexual past. In effect, while people might comment privately on the woman's past, in public she was treated with all the respect due to a woman in her position. This situation poses a conundrum if we try to account for it through Tuleños' claims that a woman's value resided primarily in the number of men who "had known her [sexually]." The fact that a woman's past could be erased once she was under a man's charge makes clear that, in practice, feminine value resided less in the number of men a woman had been with intimately than in whether she was currently under the protection—and control—of a man. This suggests also that the ideological work performed by Tuleños' moral evaluations of women was not, as one might expect, to excavate women's pasts, sift good from bad women, and distribute status rewards and punishments accordingly. On the contrary, Tuleños were singularly blind to women's pasts, for their goal was to bring and keep women under men's control in the present. A bad woman would therefore always have a chance to occupy a respectable place in the sanctioned organization of gender. For it was only through people's willful forgetfulness that they could keep bad women enticed by the rewards of the house and thus tame the contradictory dynamics that continuously destabilized El Tule's gender order.

CONCLUSIONS

In a conversation I had with Justino (Luisa's father) about Tuleño women's sexual behavior he complained that "...women no longer 'give themselves their place' "[no se dan su lugar], an expression that means that they don't respect themselves and are therefore unable to earn others' respect. Like so many men I spoke with, what Justino did not seem to understand is what every woman in El Tule knew so well: that women could not necessarily construct the life they wanted by "giving themselves their place." Women gleaned this knowledge, as subordinate social actors often do, as they dealt with contradictions in their lives—in this case, between women's conjugal expectations and the actual economic strains and instabilities of their conjugal relations and between norms of female propriety and "improper" desires. As these contradictions denaturalized local notions of proper womanhood (and manhood), they opened a space for contestation. As the cases of women such as Luisa and Maribel have shown, young and older women alike partially penetrated (Willis 1977) the workings of dominant gender ideologies and used this knowledge to carve a space for sexual pleasure and choice of conjugal mate across various stages of their life cycle. Tuleño women, that is, recognized that statuses were not fixed, but were rather historical and negotiable, and they used this knowledge to maneuver themselves into relationships and positions they desired beyond the legitimately allowed first and only relationship of their youth.

I have chosen to characterize women's practices as expressions of "agency" rather than "resistance" or "power" in order to draw attention to the complex subjectivities that I see expressed in women's sexual practices. That is, although Tuleño women could be seen as resisting men's power as sanctioned by dominant norms as well as exerting power from below, this kind of perspective has tended to either elide or grant subjects too much autonomy (Ortner 1995, Abu-Lughod 1990). Instead, I have approached women's transgressions as culturally constructed, historically emergent "forms of [structurally] embedded agency" through which women expressed a "sense of self as an authorized social being" (Ortner 1996, 17). This approach has shown how gender contradictions opened a space for women to develop their own choices and desires and act upon them in ways that were not independent of dominant norms, but were also not merely reactive. The notion of agency invoked here involves a conception of women's motivations that goes beyond instrumental objectives: the status women wanted and *who* they wanted are inseparable in this formulation. In recognizing men's collusion with women's transgressions, and women's efforts to actively encourage others' sexual misconducts, I have also recognized that women's agentive practices were enabled by multiple social dynamics, and not simply by individual will.

A feminist preoccupation with women's agentive practices also entails searching for possibilities for transformation. Thus I wonder whether such possibilities could inhere in women's success in achieving forms of social mobility that disrupted established gender norms. For while women's transgressions can be seen as incorporated into and thus part of the dominant ideology (Foucault 1978), a critical understanding of the ideological work of gender (Poovey 1988) points us to the tensions within this ideology as potential loci of change. The existence of a critical mass of women involved in these practices matters here, as it means that these were not simply individual responses but rather widespread social practices. As such, they had the potential to disrupt men's control over women's bodies and the criteria by which to assess their social value. In this regard, it is important that Tuleño women's transgressions not only destabilized the symbolic boundaries between good and bad women in community eyes—as Justino's statement above indicates—but also that their pervasiveness rendered these very *material* social positions, as defined by the instability of their boundaries, part of the landscape of "the normal" in the village. Normalization moderated the social dangers of women's transgressions while expanding "the possibilities, opportunities, and permissions for [women's] pleasure" (Vance 1992, xvii). I take Tuleño women's efforts to affirm their rights to their own bodies and erotic identities to be important in its own right. Whether the pervasiveness of these practices will lead to a *recognition* of these rights as legitimate remains to be seen.

As I envision the potential of women's transgressive practices I want to be mindful not to romanticize the possibilities for gender transformation they may open. Here it is important to recognize that even legitimized rights to erotic pleasure and conjugal choice throughout the life cycle does not mean that emancipatory gender arrangements are on the horizon. It is also necessary to recognize that despite women's success in navigating against the current of local sexual/moral norms, their orientation toward the status of woman of the house was scripted by the dominant ideology. As such, it was key to the reproduction of the local patriarchy. This orientation, which was embedded in a multi-dynamic system of power that exceeded male–female relations, was also one that pitted women in fierce competition and conflict with one another. To the extent that these features of women's practices reproduced women's gender oppression, it is appropriate to speak of El Tule's local patriarchy as a form of gender hegemony (Ortner 1990). But it is also important to remember that, inasmuch as norms are sedimented in the body and reproduced within a determinate historical praxis, challenges to these norms can only arise from specific contradictions, making unevenness an inevitable feature of change.

I have argued elsewhere (Montoya 2001b) that the force of hegemonic norms notwithstanding, Tuleño women's transgressions went beyond reproducing gender arrangements, for they enabled women to contest and redefine dominant understandings of good women. This meant that women increasingly aspired to the status of the house for the economic support and respect that it ideally entailed, but not because they felt it legitimately prescribed how they were to live their lives. Thus, if no less than elsewhere, El Tule's gender hegemonic and counter-hegemonic projects are mutually constituted (O'Hanlon 1988), it is possible to see women's sexual transgressions not as simply reactive instances of resistance but as everyday forms of gender formation latent with possibilities for change.

NOTES

I would like to thank Janise Hurtig and Lessie Jo Frazier for their careful comments on earlier drafts. Thanks also to Susan Paulson and Webb Keane. The historical and ethnographic fieldwork on which this paper is based was funded by grants from the Social Science Research Council, National Science Foundation, Wenner-Gren Foundation for Anthropological Research, the Horace H. Rackham School of Graduate Studies at the University of Michigan, and the Faculty Research and Creative Activities Fund at Western Michigan University. All proper nouns in the paper are pseudonyms.

1. I carried out fieldwork in El Tule from June 1992 to August 1993, in March 1995, in July 1997, and in September and October 2000.
2. While most people in El Tule eloped rather than married religiously or by law, they nonetheless regarded their conjugal mates as their spouses.
3. This literature is vast. For Nicaragua, see especially Bayard de Volo 2000.
4. Behar 1993, Lamas (Chapter 11, this volume), and Zavella 1997 are especially notable exceptions.
5. See McNay 2000 for a review and discussion of feminist theories that posit subjectivity in these terms.
6. For especially useful discussions on decentering categories of public/private (or public/domestic), see Cubitt and Greenslade 1997, Diaz-Barriga 1998, 2000, and Stephen 1997.
7. An important exception were women's occasional empathetic statements about other women's transgressions and their refusal to morally judge these women. See Montoya 2001b.
8. Same-sex sexual relations between self-identified heterosexual men existed, but were not visible enough to affect the sexual economy. See Lancaster 1992 for a discussion of homosexuality in Managua.
9. The proportion of single mothers, most of whom remained in the village living with their families, was small by Nicaraguan standards—usually less than 10 percent at any one time since the early 1980s.
10. Some villagers claimed that the introduction of contraceptives in the 1980s was also a factor facilitating young women's sexual experimentation.
11. On virginity in Nicaragua, compare Montenegro 2000, 67–71, and Lancaster 1992, 30.
12. See also Browner and Lewin 1982 on Calefia women's strategies.
13. For an interpretation of the blurred boundaries between house and street among Nicaraguan prostitutes, see Adahl 2001.

REFERENCES

Abu-Lughod, Lila. 1990. "The Romance of Resistance: Tracing Transformations of Power through Bedouin women." *American Ethnologist* 17(1): 41–55.

Adahl, Susanne. 2001. "Creating Safety in an Uncertain World: Shifting Concepts of 'Calle' and 'Casa' in the World of Prostitution in Managua." In *Lo público y lo privado: Género en América Latina,* edited by Maria Clara Medina. Gotemburgo: Red Haina/Instituto IberoAmericano, Universidad de Gotemburgo.

Bayard de Volo, Lorraine. 2000. *Mothers of Heroes and Martyrs: Gender Identity Politics in Nicaragua, 1979–1999*. Baltimore: Johns Hopkins University Press.

Behar, Ruth. 1993. *Translated Woman: Crossing the Border with Esperanza's Story*. Boston: Beacon Press.

Browner, Carole and Ellen Lewin. 1982. "Female Altruism Reconsidered: The Virgin Mary as Economic Woman." *American Ethnologist* 9(1): 61–76.

Cubitt, Tessa and Helen Greenslade. 1997. "Public and Private Spheres: The End of Dichotomy." In *Gender Politics in Latin America: Debates in Theory and Practice*, edited by Elizabeth Dore. New York: Monthly Review Press.

Diaz-Barriga, Miguel. 1998. "Beyond the Domestic and the Public: *Colonas* Participation in Urban Movements in Mexico City." In *Cultures of Politics, Politics of Cultures: Re-envisioning Latin American Social Movements*, edited by Sonia Alvarez et al. Boulder: Westview Press.

——. 2000. "The Domestic/Public in Mexico City: Notes on Theory, Social Movements, and the Essentializations of Everyday Life." In *Gender Matters: Rereading Michelle Z. Rosaldo*, edited by Alejandro Lugo and Bill Maurer. Ann Arbor: University of Michigan Press.

Foucault, Michel. 1978. *The History of Sexuality. Volume I: An Introduction*. Trans. Robert Hurley. New York: Pantheon.

Kandiyoti, Deniz. 1991. "Bargaining with Patriarchy." In *The Social Construction of Gender*, edited by Judith Lorbel and Susan Farrell. Newbury Park, C.A.: Sage.

Lancaster, Roger. 1992. *Life is Hard: Machismo, Danger, and the Intimacy of Power in Nicaragua*. Los Angeles: University of California Press.

Massey, Doreen. 1994. *Space, Place, and Gender*. Minneapolis: University of Minnesota Press.

McNay, Lois. 2000. *Gender and Agency: Reconfiguring the Subject in Feminist and Social Theory*. London: Polity Press.

Montenegro, Sofia. 2000. *La cultura sexual en Nicaragua*. Managua: Centro de Investigaciones de la Comunicación.

Montoya, Rosario. forthcoming. "House, Street, Collective: Revolutionary Geographies and Gender Transformation in Nicaragua, 1979–1999." *Latin American Research Review* 38(2).

——. 2001b. "False Promises of 'the house': Gender Contradictions and Women's Agency in a Nicaraguan Village" (under review).

—— n.d. *Ambivalent Revolutionaries: Exemplarity and Contradiction in a Sandinista Model Village, Nicaragua, 1979–1999* (in progress).

O'Hanlon, Rosalind. 1988. "Recovering the Subject: Subaltern Studies and Histories of Resistance in Colonial South Asia." *Modern Asian Studies* 22(1): 189–224.

Ortner, Sherry B. 1990. "Gender Hegemonies." *Cultural Critique* 14 (Winter): 35–80.

——. 1995. "Resistance and the Problem of Ethnographic Refusal." *Comparative Studies in Society and History* 37(1): 173–93.

——. 1996. "Making Gender: Toward a Feminist, Minority, Postcolonial, Subaltern, etc., Theory of Practice." In Sherry Ortner, *Making Gender: The Politics and Erotics of Culture*. Boston: Beacon Press.

Poovey, Mary. 1988. *Uneven Developments.* Chicago: University of Chicago Press.

Ricoeur, Paul. 1994. "Imagination in Discourse and Action." In *Rethinking Imagination,* edited by Gillian Robertson and John Rundell. London: Routledge.

Stephen, Lynn. 1997. *Women and Social Movements in Latin America: Power from Below.* Austin: University of Texas Press.

Vance, Carole S., ed. 1992. *Pleasure and Danger: Exploring Female Sexuality.* London: Pandora.

Willis, Paul. 1977. *Learning to Labour: How Working Class Kids Get Working Class Jobs.* New York: Columbia University Press.

Zavella, Patricia. 1997. "'Playing with Fire': The Gendered Construction of Chicana/Mexicana Sexuality." In *The Gender/Sexuality Reader: Culture, History, Political Economy,* edited by Roger Lancaster and Micaela di Leonardo. New York: Routledge.

2. Gender's Place in Reproducing and Challenging Institutions and Ideologies ∽

4. Forging Democracy and Locality ✎

Democratization, Mental Health, and Reparations in Chile

Lessie Jo Frazier[1]

The chant from a 1990 human rights demonstration I participated in highlighted a litany of places in Chile where the remains of the victims of the military dictatorship had been recovered: Mulchén, Lonquen, Laja, Pisagua, and Colina. We marched through the streets of the capital, Santiago, proclaiming, "It wasn't a war, it was a massacre, all were assassinated," and finally, "They spilled the blood—now they want to erase their guilt. There will be neither pardon nor forgetting in the earth—Pinochet is guilty. Justice and punishment for all of the guilty," indexing the struggle over whose history of the dictatorship would achieve credibility and which governing regimes would be considered legitimate. Upon the conclusion of its 1973 to 1990 rule, the Chilean military claimed to have won a civil war fought against the forces of global communism, while the human rights movement referred to a long, national history of the repression of the Chilean people. The exhumed bodies became artifacts of this struggle over history as forensic anthropologists traced the stories of torture and execution encoded on the corpses. Each additional mass gravesite mapped out the topography of state terror.

The first grave excavated in the time of regime transition occurred in June of 1990, in the northernmost desert province. The excavation riveted the nation's attention: Not only was it the first mass grave to be excavated under the new civilian government, but also the lack of moisture and salinity of the sands had kept the bodies in a state of near-perfect preservation, producing graphic evidence of state violence heretofore denied. But the grave's significance had more profound roots, as I discovered when I went to the North and learned of incidents of state violence spanning even further back than the beginning of the twentieth century, such that the northern region had come to be regarded as a place largely constituted through state violence (Frazier 1998).

With the excavation of the mass grave, the newly governing civilian coalition of center to left parties seized the opportunity to show its commitment to facing the legacies of the past. The government chose the regional capital, Iquique, as the site for a pilot reparation program for victims of the dictatorship: the Mental Health and Human Rights project. The project represented a key attempt by a democratizing state to incorporate into state policy aspects of the mental health and human rights movement that had spanned the Americas in response to the rise of military repression in the 1970s and 1980s and in turn to provide a model for other democratizing states around the world. This chapter presents a feminist analysis of the state's reparation project in the context of northern Chilean landscapes as linked with national and transnational processes.

Landscape, as an analytic category, foregrounds the ways perspective shapes place; it encompasses sites of struggle through which power relations are practiced, challenged, and reconfigured, simultaneously shaping contending understandings of place. While the study of landscape entails the geography of particular ways of seeing (Cosgrove 1985), an analysis of political landscapes must also account for technologies of state violence that do not necessarily leave visible scars, but that do tear the connective tissues of minds, bodies, and collectivities. Feminism's insistence on reading between the lines of power—as dominant narratives often leave women's lives unscripted—provides an analytic method for exposing the fissures, gaps, and absences in landscapes marked by state violence. Yet, feminist geographers have insisted that our analyses move beyond placing women on the fields of power to questioning the framing of ideas about place and power (Women and Geography Study Group 1997).

Applying this methodology to the Chilean case requires that we go beyond an analysis based on naturalized dichotomies of surface and reality in considering the excavation of mass graves, though surface and reality are modernist images central to the rhetorical fields constructed around landscapes of state violence. Under military rule, landscapes of violence worked, in part, through the creation of a landscape of order by scripting violence onto that landscape as known yet unseen, an unrelenting yet deniable threat to some and a secure space for others. Thus, landscape features and the sentiments attached to those features took on a complex lexicon that referenced different meanings depending upon political and class positions.

Local human rights activists struggled to forge a new kind of relationship to the state through asserting the particular importance of their local political landscapes.

Northern Chile, and more specifically the regional capital, Iquique, provides a case study of the gendered processes by which a military regime reshaped urban landscapes and the possibilities for locality, a form of collective subjectivity organized around a sense of particular place. It also exemplifies how a collective movement engaged in a key process of democratization, namely, the renegotiation of the relationship between the state and sectors of the populace, a relationship constituted in the junctures between forms of nation and locality. In exploring the connections between gender, place, and political power, I heed Gavin Smith's (1998) critique of the current predominance of spatial metaphors in social analysis (replacing metaphors of production) as revolving around notions of inclusion and exclusion, center and margin, such that this use of the spatial tends to emphasize agency as individual resistance to structures rather than see action in relation to and emergent from social relations of power.

I depict the features of local historical geography in the pursuit of an anti-essentialist depiction of place and locality recognizing not "authentic places," but rather nexuses of social relations grounded in struggle (Massey 1994, 121). By conducting an ethnography of landscape and place as gendered terrains of power, I seek to partially explain local struggles for mental health in the context of ambiguous nation-state processes of regime transition and Chile's neoliberal positioning within global economies. This volume's core concept, *desalambrar,* foregrounds the use of local knowledge to dismantle structures of power. In this ethnographic case study, I show the gendered processes through which local actors engage with and attempt to negotiate dominant boundaries of place and in this process, create—in feminist geographer Gillian Rose's (1993) term—paradoxical places, whose complex array of relations and meanings exceed the bounds of singular frameworks, where the subject can be in more than one place at once, demanding a more multidimensional analysis of the emergent relation between power and place.

LANDSCAPES OF STATE VIOLENCE IN CHILE
FROM CAPITAL TO FRONTIER

Human rights groups insisted that mass grave sites—covering the entire country and dating from various points during the dictatorship—were evidence of the systematic and institutionalized nature of human rights abuses under the military regime. They called for: (1) the recognition that torture was systematic and institutionalized, and not the excesses of a few officers; (2) the understanding that torture left lasting physical and psychological scars on the Chilean people (thus survivors of torture and indirect victims should be recognized); and (3) the government's recognition of international conventions against torture. These resolutions arose largely in response to the government Truth and Reconciliation Commission's drastically limited scope of inquiry: It had gathered documentation only for the cases of victims who died, ignoring survivors of torture; pursuit of justice or even complete access to information from the military in specific cases was beyond their purview; and family members of the missing or deceased could file individual court cases. Activists recognized that the repressive apparatuses institutionalized over seventeen years of dictatorship were largely left intact under the transition to formal democracy. To a certain extent, the archaeology of state violence in the excavations of the mass graves had displaced the political need for an archaeology of violence as manifested in living bodies. Torture as an ongoing process in the minds and bodies of survivors, not to mention cases of brutality since transition, became the hidden subtext of a "truth" defined by the politically possible (Vidal 2000). A humanist argument about the social body as the object of state terror was subsumed under organic metaphors of the nation as a singular, homogenous body (in this case, mestizo and neoliberal), disturbingly reminiscent of the military's use of bodily metaphors to justify the excision of subversive, infected elements from the national body (Frazier, forthcoming).

The Chilean military had operated explicitly from the Cold War principles of the National Security Doctrine, which relocated threats to the nation from its borders with other states to internal affairs where subversive danger lurked, especially among seemingly benign places and people. Cold War paranoia animated the shadows of the national landscape, finding the stain of ideology in each crevice. Thus, the military invested great energy in carving Chile into new administrative units, each with a regional command structure; it also closed down civil society and invaded domestic spaces. In response, so-called new social movements emerged to contest military rule. Scholars have analyzed these movements as being gendered female for several reasons: the predominant role played by women activists; the movements' emphases on domestic concerns such as collective kitchen purchasing cooperatives, neighborhood construction, and agentive claims based on ideas of motherhood, epitomized by the Mothers of the Disappeared; and lastly, because their politics could be constituted as a lack in the sense of the absence of normative political agency in the form of a political party and corporativist union access to the state. Under the hypermasculinization of the state under military control, human rights discourses accrued even more heavily gendered characteristics, a dynamic that continued in the shift to civilian rule.

The concerns voiced by the human rights community could not outweigh the enormous pressures on the new government to limit the truth commission, pressures revealed in promilitary sectors' approval of the report: "[the report] will not be new, because everyone knows that there were war councils—which are established by legislation—executions, and human rights violations..."[2] simultaneously occulting information about violence and audaciously, matter-of-factly claiming its routineness. This unsettling rhetoric that at once brags and denies, continued the psychological warfare deployed as a very intentional feature of state violence in which a veneer of calm and order accompanied arbitrary and random assaults on the populace—that is, the crafting of horror through the perpetuation and violation of modernist delimitations of the surface and the subterranean in landscapes of state violence.

The political party leaders of the new civilian government claimed that the survival of political democracy in Chile depended on a project of national reconciliation. The government's desire to "turn the page of history" necessitated a recognition and then transcendence of ideological conflicts marked as aberrations in an otherwise unified national history. Thus, when the civilian government advocated truth and reconciliation, it erased the issue of justice, delimited the nature of truth to tightly circumscribed investigations, and confined reconciliation to forgiveness. Forgiveness meant that those subjected to state repression would not harbor malice toward the military and the military and its supporters would deal respectfully with those people previously labeled as subversives.

In sum, public discourses of both promilitary and democratic sectors shared notions of the landscapes forged under military rule as both known and obscured, exceeding simple distinctions between the surface and the subterranean. In both cases the Chilean landscape had been feminized, in the sense of having been passively inscribed upon by the masculine agency of the military, and now subject to a knowing, scientific, and rational gaze in the contestations of those (coded male) political sectors vying for political power, including the right to name, label, and interpret that historical geography. The gendering of place and actors thus became central to the allocation of historical agency and the power to shape Chile's future. Human rights movements played on the gendered logic of military rule by casting themselves as virtuous actors defending universal principles (as opposed to sectarian political subversives) and contested the military's rendition of the national landscape by repopulating it with the embodied memories of those the military had expunged from the national body.

HISTORICAL LANDSCAPES OF NORTHERN CHILE

When I was preparing to travel North, people in Santiago often warned me that, "Iquique was General Pinochet's 'pet' city." Iquique's development via a free trade zone was to become an international exemplar of the benevolence of military rule and neoliberal economics. Paved, watered, urbanized, and blessed with a bounty of imported consumer goods, this desert port served, for both political and sentimental reasons, as General Pinochet's retreat when assailed by the demands of democratic transition. Yet Iquique's streets and neighborhoods bear the scars of the dictator's attentions. Home of key garrisons and the stronghold of the northern frontier, during the dictatorship, military residential compounds were strategically built around and within neighborhoods once famed as "the cradle of the Chilean Left" as the stronghold of Chile's first mass unions and political parties. Streets and neighborhoods were renamed after military officers and dates of military battles in an effort to erase the progressive legacy of this mining boomtown. In 1973, political prisoners had been concentrated in a nearby isolated port in a military prison camp of explicitly fascist design. This port had actually served as a place of detention several times over the course of the twentieth century (Frazier 1998). With the discovery of the mass grave in 1990, the North again became a key national marker of the search for truth and justice. In Iquique, associations of ex-political prisoners and families of the disappeared and executed demanded reparations, including social services for the victims of state terror. Subsequently the state designated Iquique as the site of the U.S. Agency for International Development (US-AID) funded pilot health program for living victims of the dictatorship. The program's success propelled its implementation throughout the country and its eventual adoption by the Ministry of Health. From the perspectives of northern human rights groups, the incorporation of human rights and health issues into a state project represents an enormous achievement for those who had contested state terror. Local human rights groups demanded reparation in the names of all who suffered state violence, and insisted on public recognition of their leadership of the pilot program by virtue of the national importance of their local histories of struggle.

The northernmost frontier province has long played a particularly significant role in Chilean state formation. Late nineteenth- and early twentieth-century struggles over nitrate wealth in the desert and over control of nitrate workers shaped the social geography of the region and its place in the nation. The military particularly valued the region for its strategic importance as a frontier and its consequent prominence in military history.

Thus the military dictatorship attempted to dismantle the Leftist component of regional history and assert itself as the region's guardian and benefactor.

In the nineteenth century, nitrate mines sprang up throughout the northern desert: These were sites of violent labor–management confrontations in an industry that provided the nation with its most significant source of revenue. Whereas the mining camps had served a key role in the world fertilizer and munitions industries, by the mid-twentieth century they had become ghost towns littering the landscape, rendered useless by the synthetic processing of nitrate. The 1929 demise of the nitrate industry marked a period of regional economic depression and the people of the region declared themselves abandoned by the state (Frazier 1998). The tenacity with which local residents continue to assert their region's importance to the nation testifies to "the importance of place when capitalism moves on," especially in light of "capitalism's successes in constructing places" (Smith 1999, 164) as well as to the ways in which local processes exceed the bounds of capitalist place-making. The North is, indeed, a space made and unmade by flows (Kasakoff 1999) of capital and people. The residues of these boom-and-bust cycles (perceived as such by people in the region) have sedimented into a remarkably stubborn, though often contradictory, sense of place.

In the 1980s, the region rose again to national prominence. The dictatorship's regional neoliberal economic policies included a new opening up to the world via a free trade zone, the fencing off of the coastline for the private benefit of the fish meal industry, and a cluster bomb plant. The military government dismantled the old nitrate camps and sold pieces of roofs, homes, stores, and industrial plants for scrap metal. The dismantling of the camps deprived former residents the possibility of continuing their periodic pilgrimages back to these frontier mining outposts to recall their lives spent working there, though they do continue to visit the one remaining camp that stands for all of their lost homes.

Given the region's economic and political history, the geography of nonelite protest became a critical interrogation of this particular landscape of state violence. As local scholar Guillermo Ross-Murray has observed (personal communication), Iquique's urban geography differed from many Latin American cities and from the capital city of Santiago. The cathedral is not located in the main plaza; instead it sits off to the side of the secondary town square that, as a site for political meetings and rallies of local and national import in front of city hall, served as the focal point for working-class communities. The main town square is Plaza Prat, named for the martyr of a naval battle, and is the site of military processions and elite promenades each Sunday, framed by an ornate neoclassical theater built of Oregon pine and adorned with sculptures of the four seasons—a curious feature in this desert climate. English nitrate barons built the landmark theater, but it was restored by General Pinochet's appointed city mayor and used by the military and its supporters to promote high (by their definition, European) culture as a social space for regional elites. Other buildings on this main town square include an elegant hotel, banks, and private clubs founded by the European capitalists of the nitrate boom years before 1929. In sum, the urban geography of Iquique refers more to turn-of-the-century, British-led investment than to Spanish colonial iconography. Thus the main plaza became associated first with foreign capital and then with displays of military/national import and the promenading of elites, while the secondary plaza became associated with the more rowdy affairs of nonelite sectors, including political contests.

During the dictatorship, protests against military rule usually began near this secondary plaza, at the Cathedral, and ended at a local school in the center of the city that had been the site of a massacre of nitrate workers in 1907. A small monument here (circa 1957) provided an anchor for subsequent urban protests and the anniversary of the massacre became a day of protest, especially during the military dictatorship, linking generations of suffering and struggle; demonstrators frequently incorporated into their protest songs from a cantata about this massacre composed by a local son during the years before the military coup. In addition to urban protests and in spite of the logistical difficulties of travel in this militarily-occupied region, human rights protesters began to make annual pilgrimages to the nearby former detention camp to commemorate the experiences of ex-political prisoners and to search for the graves of the executed and disappeared. The human rights movement in the North thus drew together regional histories of struggle to challenge the legitimacy of military rule and its constitution of the northern landscape as predominantly made up of sites of military history and dangers to the nation-state resulting from its frontiers and from the legacies of nonelite political mobilizations. While limiting its challenge to nationalist definitions of landscape, activists made claims to a locality by playing on those nationally significant features to assert their particular placement in the national landscape and thus challenged the authoritarian construction of national landscapes by renegotiating the interplay between local and capital gazes.

PILOTING REPARATION ON THE FRONTIER: THE MENTAL HEALTH AND HUMAN RIGHTS PROGRAM

National attention turned to the North in June of 1990 with the startling excavation of the mass grave from 1973. Members of nongovernmental

mental health organizations in Santiago and the newly inaugurated civilian government had been discussing a treatment program for victims of the dictatorship as part of a reparations initiative. Local human rights activists, including ex-political prisoners and their families and committed medical personnel, began to pressure the government for just such a program. The government seized the political moment and started a pilot project immediately in Iquique. Gradually, the program expanded to cover other cities and regions. The project became known as PRAIS—*Programa de Reparación y Atención Integral de Salud y Derechos Humanos* (the Human Rights and Integrated Health Service Reparation Program).

Funding for the project's first three years came from the Agency for International Development (AID). Framing the mental health project under the rubric of national development echoed the civilian government's project of "cultural modernization" (Brunner et al. 1989) designed to mold Chile to the form of a liberal democracy with a neoliberal economy. The state's underlying supposition entailed an understanding of political violence as the product of irrational ideological passions that could be tempered under the Liberal framework of politics that erased conflicts of interest—economic, political, social, or cultural—in favor of a political system comprised of citizens defined as undifferentiated, individual, rational subjects; "cultural modernization" would shape these citizens by educating Chileans away from sectarian solidarity. Inasmuch as the PRAIS program provided psychological therapy for a particularly politicized sector of society, the program significantly corresponded to this cultural modernization. Furthermore, civilian political leaders framed the transition from military to civilian governance as reconciliation, which they understood as a project of national healing in which forgiveness would suture together previous antagonists to serve the goal of national integration. The Mental Health and Human Rights program both echoed and constituted this framing of political process (institutional/structural abuse) as psychological healing (personal problem).

The PRAIS program served families of the executed and disappeared, returned exiles, and ex-political prisoners (many of whom are also returned exiles or suffered internal exile) and their immediate families, providing basic primary care as well as access to specialists. Resources were limited, and patients often had to go to great lengths to obtain appointments, with complicated procedures to obtain referrals to higher levels of care. In Iquique, between September 1990 and December 1993, the project attended to 798 patients. About 77 of these were members of 36 families affected by situations of execution or disappearance. On a national scale, PRAIS estimated that it had reached 37.8 percent of potential patients from such families (Domínguez et al. 1994). As I discuss below, the state's concern with reaching this particular segment of the population—families (presumably but not necessarily) indirectly involved with political conflict through the death of a politically militant relative—is indicative of the way in which PRAIS operated with an ideal, and gendered, model of both the patients and the process of recovery.

The program, as with other reparation policies such as cash payments, educational scholarships, and state-financed reburials of those killed by the military, privileged family members—especially parents, spouses (almost exclusively wives), or children—of individuals executed or disappeared by the military regime (individuals gendered male by their political actions). Consequently, the ideal process of recovery was that of mourning in which the family (as the foundational unit of civil society) comes to terms with its loss and then moves on.[3] As with the Truth and Reconciliation Commission report, this ideal model failed to account for surviving individuals who were detained, tortured, and exiled, or for the direct participation and repression of families of the disappeared in resisting the dictatorship. Many relatives of disappeared or executed people became active in the human rights movements and political mobilizations against the dictatorship and, consequently, the state arrested, interrogated, harassed, or even disappeared them as punishment for their involvement. In other words, many family members of the disappeared had been directly subjected to state repression as well. However, this component of their historical experience was ignored and the civilian government portrayed the families of the disappeared as ideal-typical victims, that is, passive and thus innocent objects of actions beyond their control. Defining human rights as pertaining to people subjected to unprovoked suffering was a critical tactic for a civilian government operating without the consensus of the populace (44 percent of whom had still voted for the military faction) that prior state violence had not been justified by the danger of internal, subversive agents (who therefore deserved the consequences). Thus, despite their universalist rhetoric, human rights policies in Chile, as in most parts of the world, place a premium on measures of innocence and victimhood. In sum, the rhetorical field of human rights discourses constitute as agentive—in the Chilean case at least, masculine—the military perpetrators and the civilian political leaders who advocate on the victims' behalf, while casting as passive—feminine—the family members of the disappeared and executed who were exemplary members of the patient population (as opposed to the ex-political prisoners).

Given this ideal type of patient as family member and the ideal type of therapeutic process as mourning a deceased loved one, it is curious that the pilot program for the entire country took place in the North. This is a region where the largest and most organized sector of the human rights movement at the time of transition to civilian rule was the Association of Ex-political Prisoners, who embodied the antithesis of the state's ideal model in being mostly middle-aged men tainted by their direct participation in politics. Furthermore, most nongovernmental organizations working in mental health and human rights operated primarily in the capital, Santiago. In a country spanning over 4,200 kilometers, the logistics of governmental and nongovernmental programs are no small matter. With almost one quarter of the population concentrated in the central, metropolitan region, most programs have few ambitions beyond the capital. Yet, in a country historically centralized in economic and political power as well as administrative bureaucracy, frontiers have played a key role in state formation. The political process of regime transition from military to civilian rule did not prove exceptional in this regard; the pilot for the most far-reaching reparation project was centered in the North, one of the most intensely repressed areas of Chile during military rule as national security interests in this remote corner of the country had become vital to the military's notion of itself as the guarantor of national honor and the integrity of the Chilean nation-state. Northern social movements, including the Association of Ex-political Prisoners, had successfully disseminated their counter-rendition of local historical geography (as organized struggle against a long history of state violence) throughout Chile.

Thus the pivotal role played by the regional ex-political prisoners' association proved to be perhaps the most unlikely feature of this reparations initiative, underlining a core problem not only in the reparations project, but also in the very model of democratization itself. The civilian leadership's timetable for transition from military to civilian rule factored in a period of national healing culminating in reconciliation understood as forgiveness. Chile was the first place where reconciliation served as the primary objective of a truth commission, a model soon exported to other countries, such as South Africa. As noted previously, the Chilean Commission on Truth and Reconciliation only documented cases resulting in death. Forgiveness for the deaths of those gone forever seemed a more feasible objective that confronting the ongoing coexistence of tortured and torturers. Other complex issues remained unresolved, including issues of property disputes and reparation for economic damages. Many military officials and supporters profited economically from their involvement with the military regime, just as many other sectors of the population had suffered great losses of jobs and property. Furthermore, many ex-political prisoners have been denied the right to vote or obtain a passport because they technically have criminal records from the courts-martial, even though these crimes were presumably covered under the military's own amnesty law. Lacking the political will, or power in some cases, to pursue such issues, the civilian government invested itself heavily in delimiting the extent of the official investigation of the, in their term, truth of human rights violations. As discussed below, Iquique's resistance to the easy incorporation of the PRAIS program into the bureaucracy of the Ministry of Health highlights the contradictions inherent in this pragmatic model of regime transition as healing.

The state's discomfort in tracing the history of repression as inscribed on the living (a counterpart to the forensic analysis of those found in the grave) shows itself in the disagreements between staff and patients about the connection between current health complaints of a generally aging population and the physical and psychological traumas experienced over the last 20 years. The ex-political prisoners, predominantly men, attributed diabetes, stress-related disorders, cancer, alcoholism, and other common illnesses to physical abuse and torture in the prison camp and detention centers. On a national level, PRAIS estimates a 28.4 percent relation between diagnosis and the repressive situation under which patients lived. However, when considering mental illness apart from other complaints, the PRAIS data analysts in Santiago found a 48 percent connection between mental illness and individual history of repression with a 12.8 percent relation for physical problems (Domínguez et al. 1994). Let us consider this distinction between mental and physical illnesses.

Overall, 47.5 percent of diagnoses were for mental disorders. The predominance of mental as opposed to physical diagnoses was particularly notable among women. This difference most likely reflects the tendency—alluded to in the PRAIS report and noted in my interviews with PRAIS staff—of many men to be much less willing to engage in mental health treatment and more likely to consult PRAIS solely for physical complaints. These men's investment in the very distinction between so-called physical and mental ailments points to the gendering of these categories. On the one hand, some men's sense of themselves as agents (heavily invested in their sense of their own masculinity) seemed compromised by notions of psychological weakness. This was particularly evident in their critique of the program officers' patronizing attitudes toward them, and of the top-down and infantilizing structure of the program in general. In effect, the ex-political prisoners rejected the state's assertion of its paternalisic relationship to this sector of the populace by not soliciting these services. Furthermore, the ex-political prisoners' fear of subjecting themselves to psychological treatment was not unreasonable given the predominance of psychological terror

(Jelin 1999) in both their own interrogations as prisoners and in the general implementation of state terror they had lived under for nearly two decades. A number of former prisoners would recount proudly to me that while they had suffered great physical abuse in detention, they had not submitted psychologically to their tormentors. Given that prior context, it is little wonder they refused to engage in psychological therapy with state functionaries. By contrast, one might argue that the state *had* constructed gender-specific notions of their patients, given that the ideal patient was the wife, mother, or child of someone killed by the military state. Indeed, by PRAIS statistics, this definition of subordinate family members did characterize the majority of patients served by the mental health component of the program. This state shaping of and intervention into the family echoes the tendencies, noted by feminist scholars, of the mid-twentieth-century welfare state to intervene in the domestic realm and emasculate or render invisible local patriarchs. Feminist scholars, building on Foucault's work, have also pointed to the gendered underpinnings of psychology's naturalized object of study. These factors together suggest an overdetermination in the gendered dynamics of the mental health component of the program, and indeed in the modernist partitioning of state notions of health into mental and physical spheres.

So-called physical problems tended toward the osteomuscular system and connective tissue (6.8 percent) and the circulatory system (5.2 percent) (Domínguez et al. 1994). Though the PRAIS staff in Santiago was inclined to see some connection between these ailments and past violence, some of the PRAIS staff in Iquique were reluctant to draw conclusions connecting current symptoms and past traumas. In interviews with ex-political prisoners, I found that the medical staff's general skepticism frustrated the ex-political prisoners: a number of ex–prisoners refused to discuss those past traumas or, in some cases, even current physical symptoms with the staff.

In an interview with the general physician of PRAIS in Iquique, I asked about the medical basis for the ex-political prisoners' insistence that their current physical and mental problems stemmed from torture in the prison camp. The doctor replied that torture had not really been an issue in the camp because it operated early in the dictatorship before methods of torture were perfected, and most of the guards were mere conscripts and not trained to torture. (These points are historically mistaken.) In a interview session I had with a committee of the ex-political prisoners who were interested in putting together their own book on the camp, one of them insisted that an entire chapter should be devoted to their experiences of torture since torture was central to the fundamental purpose of the camp, namely, to force prisoners to betray or renounce their political loyalties. This result—that people did not tell the military what they wanted to hear—represented a victory for the survivors. The others emphatically agreed and recounted in detail some of their experiences of torture.

How do we understand the physician's denial of torture? Though personal conviction led this doctor to work in the program, in doing so the physician became a state functionary, in other words, the local component and agent of the state. Critics of Chile's model of regime transition have underscored the limitations of that process, which left unchallenged much of the prior state apparatus: the military, its constitution, its lower-level functionaries, its judicial system, its presence in the Congress, and most significantly, its neoliberal economic policies. Perhaps, then, the doctor's inability to listen to accounts of torture reflected the difficulty of professionally representing the same state that had perpetrated that very torture. Perhaps, as a state officer with limited resources, this doctor was unable to offer the patients the full recognition and services they demanded. Alternately, the doctor's denial of patients' own portrayals of their conditions may reflect a general distrust of their motives, a fear of being taken advantage of by some undeserving people. While we do not have access to the doctor's motivations, we can still note that the doctor did reject vehemently a claim to reparations for specific past wrongs and a challenge to the doctor's diagnostic power as a state officer. In demanding that the state, via the PRAIS staff, acknowledge the specificity of violence in the North, the ex-political prisoners reconstructed themselves as historical agents and reclaimed their integrity, not in opposition, but with a particular relation to the state and national history.

In the process of regime transition, the PRAIS program represented the most direct link between the northern human rights community and the state. It was a tenuous connection at best. After a cautious trial period, in 1994 the program was formally and permanently incorporated into the Ministry of Health. Named the program on Mental Health and Violence, it primarily addressed issues of domestic violence, rape, child abuse, and molestation, in addition to its former role in treating direct subjects of state violence perpetrated under the military regime, a component that was expected to diminish with time with the attrition of the generation that had lived through the military coup (Frazier forthcoming).

The northern program staff resolutely refused to expand their mandate beyond the realm of so-called political violence. In response to this stand, the state allowed this regional branch of the program to continue working solely with victims of so-called political violence. Meanwhile, the Association of Women Against Violence in Iquique, like feminist groups in

Santiago, had argued that there could be no real democracy in Chile as long as women were oppressed. They broke down distinctions between political violence and domestic violence to argue that violence against women *is* political. Thus, a more radical challenge to the program on Mental Health and Violence would have questioned distinctions between realms designated as the social and the political to assert the need for state policy to move beyond triage to access to justice and structural change on multiple fronts.

While accepting the state's limited definition of the political, local health care professionals rejected the proposed change in their mission; health care staff protested that they could barely meet the needs of current patients, let alone take on whole new categories of potential clients. The Ministry of Health agreed that Iquique, unlike other sites in the country, was not seeing a decrease in patient visits. The site thus still had much work to do in the recuperation of victims of the dictatorship, and could continue a bit longer with its original mandate to heal those people (understood as individuals) deemed most directly harmed by state violence—a decision implying, once again, that state violence was not structural but rather directed at individuals.

In contrast to national and local state program officials, human rights groups in Iquique viewed the program itself not as a concession granted by a (patriarchal) state, but rather as a victory that they had won through great persistence. They make specific demands of the civilian regime largely through this program since it is one of the most direct links they have to the state. Proud that their efforts in the North had served as a health-care delivery model for the rest of the country, they saw continued local demand for the program as a positive sign of the strength of their community forged in struggle. The state, in contrast, saw the continued vigor of the program as indicating that these people were still sick, trapped in the past, and had yet to meet the goals of reintegration—in other words, that they occupied a place apart from the rest of the nation. Integrating the program into the Ministry of Health and broadening its scope was a way of signaling that Chile had put its history behind it, reconciliation was complete, and government concern could move on to so-called nonpolitical social problems, exemplified by the domestic abuse of women, children, and the elderly, in a project of cultural modernization. Thus, the ongoing, fervent sense of local integrity in terms of the North's relation to the past and to the nation-state fundamentally challenged the civilian regime's policies.

I have used the term "locality" to indicate a form of collective subjectivity involving a shared sense of a particular place as a contested, elastic terrain of struggle, a great deal of which focuses on internal and external boundaries. As discussed above, forging locality creates social relations between the local and supralocal, while also producing relations between actors practicing locality. Among northern human rights activists, there are bitter divisions; however, most all agree on the importance of the program in their daily lives and claim it as fruit of their own struggles rather than to the benevolence of the state. They make constant demands of the program and charge that it has an excessively top-down, paternalistic model of health care delivery. As I found in interviews, and as alluded to in the official report, they consider it their prerogative to utilize the program selectively in terms of which health services they use, how much they reveal to the staff, and which members of their families participate in the program.

In interviews with both local and national PRAIS staff, I found that the staff, too, emphasized the distinctiveness of Iquique both as a place in general and as a pilot site for the therapeutic program in Chile. The national director acknowledged the long history of northern political activism and the intensity of state repression in the region. On the local level, health-care professionals in Iquique contrasted themselves to the program officers in Santiago. They felt that everything they knew about caring for this population they had learned on their own with little support from the administrative offices in the capital who saw them more as "a source of data"—referring to the amount of energy they were required to spend on administrative paperwork and the gathering of program statistics. Local staff also felt that officials in the capital did not fully appreciate the stress they worked under, especially their sense of isolation and harassment by other medical staff in the hospital: for example, being hailed derisively as "reds" in the hospital corridors. The local staff has been intransigent in their refusal to take on the broadened mandate of institutionalization into the Ministry of Health under the Mental Health and Violence Program.

The PRAIS reparation program had been constructed by the state as a temporary measure until the victims of political repression recovered and were reintegrated into Chilean society. In spite of the conflicts between patients and local providers, they maintained a mutually reinforcing, if not unified, front toward the central state and its claims of proprietorship over the program. Local actors, asserting themselves as historical agents, demanded and were conceded the uniqueness of the North and its particular needs.

STATE APPROPRIATION OF THE INTERNATIONAL MENTAL HEALTH AND HUMAN RIGHTS MOVEMENT

The mental health and human rights movement spanning the Americas holds exciting possibilities for transnational, national, and local struggles for justice (Hollander 1997). The movement has expanded from concerns with the individual and practice as located in the clinic to seeing patients' health as inseparable from collective health and politics. Those concerns, moreover, are grounded in clearly defined places whose historical geography brings a great deal to bear on the conceptualization of illness, treatment, and well being. Mental health has been reformulated beyond individualized pathology to include social and political conflict, as well as positive processes of growth, well being, and social connection. For the most part, Latin American mental health and human-rights activists have operated on the margins or in direct opposition to the governments in the region, most of which were under military control at one point or another during the period from the mid-1960s to the 1970s, and in the case of Chile through the 1980s. The conflicts and challenges that arose when their insights were taken up partially in a state project has been the subject of this chapter.

While many families of the executed and disappeared received one-time cash compensation and some educational grants, the PRAIS program remains the most far-reaching government initiative in offering reparations to the survivors of state repression as broadly defined. The new civilian government's emphasis on reconciliation—understood as national healing through forgiveness and the letting go of past conflicts—meant that health became not only a metaphor for politics but also the safest means of enacting a politics of reparation. This rendering has certain troubling implications. Political repression as pathology and reparation as therapy do not necessarily confront the configurations of power that made a long-lived and, among some sectors, quite popular dictatorship take hold and endure. A brief account elucidates this point.

A North American psychologist from a prominent U.S. treatment center for victims of torture visited Iquique in hopes of including the city in a report on mental health and human rights in Latin America. He had heard of Iquique's fame as a focus of particularly terrible repression. In 1993, I attended a meeting he held with the Association of Ex-Political Prisoners. At the meeting the psychologist suggested that international assistance to their group could best be mobilized with a complete account of the challenges they have faced in organizing. They began with tales of the "shameless" who siphoned off international solidarity funds during the dictatorship. They then told of the takeover, during the transition to civilian rule, of the Association of Family Members by promilitary widows who seemed to be most interested in the reparation money. The psychologist explained to the ex-political prisoners that one insight of political psychology has been to assert that *they* are not the mentally ill; rather they are people responding in a normal way to extremely adverse circumstances and it is society that is demented. Similarly, he suggested that these widows were suffering from an overidentification with their oppressors and that through therapy they would come to see the error of their ways. After the meeting, one person—a founder of the human rights movement in Iquique who had lost a child to torture and a firing squad, had seen years of organizing swept aside by the usurpers, and had been forced to abandon the organization of family members altogether—turned aside and said: "*That's crazy.*" She rejected the psychologist's dismissal of moral culpability as pathology. If politics is pathology then there are *only* victims—including the politically subversive and those whose paranoia required their destruction. This is, indeed, the implicit rationale behind government calls for forgiveness (meaning that the human rights movements should forgive the military and its civilian supporters) as the first step toward political healing defined as the subsuming of political conflict in the greater interests of the nation. In spite of this overarching rhetoric of political conflict as pathology, for pragmatic reasons the civilian regime directed all of its therapeutic policies of national reconciliation toward those people subjected to military violence.

To the civilian government's credit, the Chilean case is a rare example of state leaders implementing in any way the insights of the mental health and human rights movement. However, in doing so the state changed the political content of that therapeutic model, which had insisted that the pathology lay in authoritarian societies and not in the people subjected to violence. Given this alteration in orientation, developing a health program as reparations (instead of a political process of justice against perpetrators of human rights abuses) is particularly ironic given the Chilean state's international role as a laboratory for neoliberal policies, including the drive toward privatization of social services such as health care, a process begun by the military and accelerated under civilian leadership (Paley 2000). Thus, health as a metaphor, vehicle, or substitute for state public policy is an ironic and ambiguous indication of the postdictatorship condition in Chile and the more generalized politics of neoliberal state formation, especially in subaltern states (cf. Cousins 2000). These ambiguities congeal in the multiple meanings and uses of particular places, as the case of northern Chile illustrates so well.

PARADOXICAL PLACES IN LANDSCAPES OF STATE TERROR

Place has served as both a source of affirmation and degradation in northern Chile. On the one hand, many ex-political prisoners and families of the executed and disappeared make the annual pilgrimage to the former detention camp where they hold a ceremony at the mass grave site. A cultural ceremony is held in the old theater, the same theater in which prisoners were forced to perform for the officers and soldiers during their time in the prison camp. Acts and offerings are also made before a small shrine in the town. On the other hand, other sectors see potential in the relationship between place and commerce: The prison was auctioned off by the military, purchased by a developer, and transformed into a hotel, complete with the original graffiti on the walls as carved out by the prisoners and the bars on the doors and windows. The private management makes no attempt to hide the building's history, but uses it as a marketing ploy. Former prisoners are not allowed on the now-private premises. Even the torturers and military officers and soldiers who helped the junta carry out their mission of terror have sought to profit from joining place and health. Upon the excavation of the grave in 1990, an officer in Santiago fraudulently claimed to have all of the particulars regarding that excavation and other gravesites. Former soldiers have extorted money from families by promising to disclose other clandestine grave sites, ironically privatizing military intelligence since the military as an institution refuses to disclose information on the whereabouts of the executed and disappeared.

Through the particular shape that reparation has taken in the Chilean transition from military dictatorship to civilian rule in the context of a deepening of neoliberal economic transformations, health and place have become central to human rights struggles. In northern Chile, human rights movements insist on the historic specificity of both their needs and their relation to the state. Activists see this history as inscribed in the scars that the dictatorship left on their bodies, minds, and lands. They allege the ongoing connections between those scars and the contemporary political order. In so doing, they demand that the state facilitate their efforts not to recover from and/or get over the past and be reincorporated into a seamless national narrative, but rather to recuperate political voice and integrity.

My analysis has moved back and forth from the intersections of regional and national historical geographies considering the landscapes of state terror and political protest to the contentious forging of the public health reparation project and interpretive struggles over ex-political prisoners bodies. In this linking of landscape and the body, I draw on insights of feminist scholars into the gendering of the body politic to understand the possibilities for taking apart a landscape of state terror justified by the military's fetishizing of the national. As with so many public health policies, the mental health and human rights reparations project makes literal the metaphor of the body politic in a manner that underlines the gendered terms of the relational categories of state and citizen. This is particularly evident in the way it reinforces what political scientist Lucy Taylor (1998) has characterized as Chile's "welfare model of citizenship," which contributed to the "depoliticization of the content of citizenship." Feminist scholars have contested the power of the objectifying gaze on the body, identifying that gaze as intrinsic to the notion of landscape and rejecting the objectifying gaze to foreground embodied subjectivities. Along these lines, I have cast as my protagonists Chilean human rights activists who have struggled to reshape relations between the state and the local and even to break down sharp distinctions between those categories by seeing themselves as citizens defending the integrity of their locality and thus participating in the creation of a new, democratic state. Their efforts exemplify the process of desalambrar, of taking down old fences and redefining political landscapes by asserting a collective proprietorship.

Formulating the cultural geography of landscapes of state violence requires attention to multiple contexts of struggle. In this chapter, I have tried to build on the current concern in both anthropology and geography to show how "the interconnection between global forces and local particularity alter the relationships between identity, meaning and place" (McDowell 1994, 166). However, despite its stress on "interconnection," this model still tends to depict a unidirectional and binary process: The global is construed as the site of agency ("force") and the local as the stage for the intertwining of identity, meaning, and place. The challenge is to see how struggles cast as local impinge upon state and trans-state projects to reveal the tensions between forms of locality (whether they be grounded in a neighborhood, a city, a region, or a nation-state) and the requirements of projects that claim to transcend locality. Thus, the case of northern Chile challenges binary analyses of place focusing on inclusion and exclusion, center and margin. Northern human rights activists used locality to shape their relations to—and demands of—the nation-state, rather than as an attempt to create an autonomous realm apart from the nation-state. If the North is a paradoxical place (Rose 1993) marked by intense strife yet also nationalist fervor, then the demands human rights activists made on the Chilean state in a time of regime transition have pushed at the limits of that paradox. In making these demands, activists resisted a centralization of suffering, challenged the very foundational political models of reparation and healing, and asserted their own politics of healing by grounding illicit events in specific places, recuperating the historical memory of their significance, and forging the connection between place, health, and justice.

This case study of a particularly paradoxical place can point to possibilities for and limits to projects of democratization. Cross-disciplinary feminist theory has demanded that we push gender analyses beyond questions of women or men to consider the gendering of political fields. This impulse is even more radically articulated in the call by feminist geographers for the "decentering of gender in feminist analyses." They call on us to move beyond gender, or "women" as the central focus of feminist scholarship (Women and Geography Study Group 1997, 194), and to focus instead on "the subjective construction of material spaces" (197). In this chapter, I have taken on a fundamental problem in processes of democratization, namely, the negotiation of new patterns of interaction between the state and particular sectors of the populace, patterns of interaction that have been historically heavily gendered in their shape and dynamic. I have also heeded Chilean scholar Manuel Antonio Garreton's (1999) insistence that the study of democratization should be grounded in particular processes rather than universalist definitions and criteria. In describing the regime transition from military to civilian rule in terms of democratization, political leaders in Chile created expectations across the nation of new spaces for citizenship and locality in the resculpting of the national political landscape. Yet, this model of democratization very much remains within the conventional framework of a landscape, that is, a rendition of place organized by a unitary perspective or gaze. Local actors have pushed against the tendency of the state to organize that gaze, thus defining the possibilities for subjectivity. We saw this dynamic in the state's limiting definitions of the political to exclude issues relegated to the modernist realms of the economic (namely the continuity of the transition to neoliberal policies) or the social (in this case, most evident in programs emphasizing domestic versus political violence). In contesting that political framework, human rights activists in Chile have strained against dichotomous distinctions between the local and the national and have struggled to place justice and the integrity of their own sense of locality on the map of national, democratic politics and in that struggle have themselves participated in the drawing of that map.

NOTES

For research support, I thank the National Science Foundation, Fulbright, the Wenner-Gren Foundation, the University of Michigan, and the University of South Carolina. For comments, I thank Andrew Cousins, Mine Ener, Julie Hastings, Janise Hurtig, Alice Kasakoff, Ann Kingslover, Susan Paulson, Rosario Montoya, and colloquia participants of the University of South Carolina and University of Chicago Anthropology Departments.

1. See an earlier version of some of this material from the perspective of medical geography: Frazier and Scarpaci 1993.

2. Julio Dittborn, cited in the on-line information service CHIP (Chile Information Project) News, February 9, 10, 11, 1991. Sources were *La Nación* and *El Mercurio:* Santiago, Chile.

3. While the nongovernmental mental health and human rights movement predominantly used variations of Marxist psychoanalysis (Hollander 1997), PRAIS relied on the intervention of social workers and on cognitive-behavioralist psychologists.

REFERENCES

Agger, Inger and Sören Buus Jensen. 1996. *Trauma y cura en situaciones de terrorismo de estado.* Santiago: CESOC.

Brunner, José Joaquín, Alicia Barrios, and Carlos Catalán. 1989. *Chile: Tranformaciones culturales y modernidad.* Santiago: FLACSO.

Comisión Nacional de Verdad y Reconciliación. 1991. *Informe Rettig: Informe de la Comision Nacional de Verdad y Reconciliación, Vol 1 and 2.* Santiago: Gobierno.

Cosgrove, David. 1985. "Prospect, Perspective and the Evolution of the Landscape Idea." *Transactions of the Institute of British Geographers* 10(1): 45–62.

Cousins, Andrew. 2000. "Ideology and Biomedicine in the Palestine West Bank." Ph.D. diss. Atlanta: Emory University.

Domínguez V., Rosario et al. 1994. *Salud y derechos humanos: Una experiencia desde el sistema público Chileno, 1991–1993.* Santiago: PRAIS (Programa de Reparación y Atención Integral de Salud y Derechos Humanos, Ministerio de Salud).

Frazier, Lessie Jo. 1998. "Memory and State Violence in Chile: A Historical Ethnography of Tarapacá, Chile, 1890–1995." Ph.D. diss. Ann Arbor: University of Michigan.

———. 1999. "'Subverted Memories:' Countermourning as Political Action in Chile." In *Acts of Memory,* edited by Mieke Bal, Leo Spitzer, and Jonathan Crewe. Hanover: University Press of New England, 105–119.

———. Forthcoming. "Medicalizing State Violence, Domesticating Human Rights in Market-States." In *Violence and the Body,* edited by Arturo Aldama. Bloomington: University of Indiana Press.

Frazier, Lessie Jo and Joseph Scarpaci. 1998. "Landscapes of State Violence and the Struggle to Reclaim Community: Mental Health, and Human Rights in Iquique, Chile." In *Putting Health into Place: Making Connections in Geographical Research,* edited by Robin A. Kearns and Wilbert M. Gesler. Syracuse: Syracuse University Press, 53–74.

Garreton, Manuel Antonio. 1999. "Review of *Patterns of Democratic Transition and Consolidation.*" *Journal of Latin American Studies* 31(3): 768–769.

Hollander, Nancy Caro. 1997. *Love in a Time of Hate: Liberation Psychology in Latin America.* New Brunswick: Rutgers University Press.

Jelin, Elizabeth. 1998. "The minefields of memory." *NACLA: Report on the Americas* 32(2): 23–29.

——. 1999. "Review of *Love in a Time of Hate*." *Journal of Latin American Studies* 31(3): 786–788.

Kasakoff, Alice. 1999. "Is There a Place for Anthropology in Social Science History?" *Social Science History* 23(4): 535–560.

Lira, Elizabeth and María Isabel Castillo, eds. 1991. *Psicología de la amenaza política y del miedo*. Santiago: ILAS.

Massey, Doreen. 1994. *Space, Place and Gender*. Minneapolis: University of Minnesota Press.

McDowell, Linda. 1994. "The Transformation of Cultural Geography." In *Human Geography: Society, Space and Social Science*, edited by D. Gregory, R. Martin and G. Smith. Minneapolis: University of Minnesota Press, 146–173.

——. 1999. *Gender, Identity, & Place*. Minneapolis: Minnesota.

Paley, Julia. 2000. *Marketing Democracy*. Berkeley: University of California Press.

Rose, Gillian. 1993. *Feminism and Geography: The Limits of Geographical Knowledge*. Cambridge: Polity.

Scarpaci, Joseph L. and Lessie Jo Frazier. 1993. "State Terror: Ideology, Protest, and the Gendering of Landscapes." *Progress in Human Geography* 17: 1–21.

Schild, Verónica. 2000. "Neo-Liberalism's New Gendered Market Citizens: The 'Civilizing' Dimension of Social Programmes in Chile." *Citizenship Studies* 4(3): 275–305.

Smith, Gavin. 1999. *Confronting the Present: Towards a Politically Engaged Anthropology*. Oxford: Berg.

Taylor, Lucy. 1998. *Citizenship, Participation and Democracy*. Basingstoke: Macmillan Press.

Vidal, Hernán. 2000. *Chile: Poética de la tortura política*. Santiago: Biblioteca Setenta&3/Mosquito Comunicaciones.

Women and Geography Study Group. 1997. *Feminist Geographies: Explorations in Diversity and Difference*. Harlow: Longman.

5. "What the Strong Owe to the Weak" ⌖

Rationality, Domestic Violence, and Governmentality in Nineteenth-Century Mexico

Ana María Alonso

Examining nineteenth-century domestic violence cases from the rural town of Namiquipa, Chihuahua, Mexico, highlights the ways in which law was a site for the negotiation of gender, ethnic, and class identities and the power differences they entailed. I analyze local cases in relation to a study of broad processes of governmentality in Mexico, where the "Liberal revolution" of the mid-nineteenth century spurred attempts to ground political authority in legal-rational rather than patrimonial forms of legitimation. In order to write about domestic violence in Mexico from my social and geographic location as a "Hispanic" anthropologist at the University of Arizona, I have had to face a dilemma. How do I critique forms of gender, ethnic, and class subjection in Mexico without feeding "Anglo" stereotypes on the United States side of the border that character-ize "Hispanic" men as violent machos and women as stoic martyrs? Here, I negotiate my dilemma by putting into question the binary oppositions that underpin ethnic stereotypes as well as forms of identity constituted by bourgeois law.

The notion of bourgeois law as a neutral arbiter of conflict relies on a spatialized separation of public and private spheres concealing a contradic-tion in the constitution of identities: "By treating those who share an iden-tity as equals, law casts as unequal those who do not share that identity. At the same time, law constitutes all people as having identities of various sorts that they should be equally free to express without hurt or hindrance" (Collier, Maurer, and Suárez-Navas 1995, 2). Oppositions between reason and force, mind and body, civilized and savage, underpin the play of sameness and difference in bourgeois law and are deeply implicated in forms of gender, ethnic, and class domination and the ways these are articulated in social space. Criminal prosecution of domestic violence in nineteenth-century Mexico was part and parcel of a project of state formation that rationalized rather than undermined patriarchy. By privileging physical force as the locus of illegitimate power, Liberal discourse represented a transformation of earlier understandings of "what the strong owe to the weak." Yet Liberal law occluded the violence and inequality promoted by the governmentalization of social relations, that is, by the management of populations and the regulation of daily life by means of techniques of governance founded on reason.

THE RATIONALIZATION OF PATRIARCHY

During the mid-1850s, Liberals gained control of the state in Mexico and attempted to revolutionize the legitimate bases of authority, no longer to be the personal will of the sovereign but the public will of the people.[1] Replacing patrimonial with legal-rational forms of political legitimation such as electoral democracy involved undermining the link between personal violence and social authority; violence became viewed as a negation of the free exercise of will and reason that legitimated the liberal contract between government and people. Punishments such as mutilation, whipping or torment of any sort that deployed physical force to inscribe subjects' lack of free will on their bodies and to dramatize personal subjection to a sovereign were banned by the Mexican constitution.

The redefinition of legitimate authority in the public sphere entailed a reciprocal transformation of authority in the domestic sphere. Mid-nineteenth-century legal commentary stated that in modern times, the domestic authority of the patriarch should be more that of "a legislator or judge" than that of a sovereign (cited in Arrom 1985a, 78). Though Liberals did not question a husband's right to govern his wife, they did redefine the basis of his authority: The subjection of woman to man in marriage was to be predicated on the legal contract, the epitome of free will, reason, and modernity, at odds with physical violence, a signifier of barbarism.

Prior to the Liberal revolution of the 1850s, both church and state had condoned a husband's right to apply "moderate physical punishment" to correct his wife (see Arrom 1976). Laws on domestic violence were "tightened" with the 1845 vagrancy code that, in contrast to its predecessor, now

included wife beaters along with drunkards and gamblers among those to
be prosecuted (Arrom 1985a, 237). Yet, the definition of antisocial behavior in this code "was limited to the husband who mistreated his wife
'frequently without any motive, scandalizing the community with his conduct.' Although the new law gave the State the power to arrest and punish
violent husbands, wife abuse, as a private crime, was normally investigated
only if a woman or her relatives filed a formal complaint" (Arrom 1985a,
237). However, under "the first Mexican Criminal Code" (Arrom 1985a,
310 n110), drafted in 1871 by the Liberals, wife beating, though not
specifically distinguished as a type of offense, could be prosecuted regardless of its severity and penalized under the general categories of blows and
lesions. Moreover, for the first time in Mexico's history, a man's murder of
his adulterous wife or her lover became a criminal offense punishable by
four years in prison. A spouse who beat rather than killed a partner caught
committing adultery was also punished, though the penalty was reduced.[2]
The provisions of the 1871 Criminal Code represent only one facet of the
Liberal revolution's extension of the rationalizing reach of the modernizing
Mexican state into the domestic domain. For Mexican Liberals, the nuclear
family, rather than the individual, was both the basic unit of society and a
microcosm of the state. Liberal legislation in the 1850s deprived the
Church of the legitimate authority over much of domestic life, and placed
it in the hands of the state.

Read to couples by judges performing the ceremony of civil matrimony,
the Epistle of Melchor Ocampo,[3] included in an 1859 Liberal law regulating marriage and divorce, demonstrates how Liberal constructions of gender and sexuality were normalized in the legal rituals of everyday life.
Within "the conjugal duality," defined as a civil contract based on free will,

> the man, whose sexual endowments are principally valor and strength,
> should and will give to the woman protection, nourishment, and direction,
> treating her always as the most delicate, sensitive, and fine part of himself,
> and with the magnanimity and generous benevolence that he who is strong
> owes to she who is weak, especially when she gives herself to him and, when
> she has been confided to him by Society. The woman, whose principal
> endowments are abnegation, beauty, compassion, perspicacity and tender-
> ness, should and will give to the husband obedience, pleasure, assistance,
> comfort, and counsel, treating him always with the veneration that is owed
> to the person who supports us and defends us, and with the delicacy of she
> who does not wish to exasperate the brusque and irritable part of herself ...

Men and women can transcend their essential incompleteness, a product of gender difference, and achieve integrity only through marriage,

which allows the recovery of the other as part of the self. But the life-long exchange that the Liberal contract establishes is unequal even if the rhetorical parallelism of the Epistle represents marriage as balanced reciprocity: Protection and obedience do not have the same social and cultural value. The Epistle exhorts husbands and wives not to insult each other as "insults between spouses dishonor the one who proffers them," and condemns "abuse in deed," a gloss for wife-beating, as "it is villainous and cowardly to abuse force"; significantly, one of the impediments to marriage cited in this 1859 law is the use of violence since it negates freedom of consent. Though men's strength is still key to the symbolic construction of their social authority, its deployment against women is now deemed dishonorable by the state, construed as a threat to the stability of marriage and family at odds with liberal forms of governmentality, and hence, of society itself.

Changes in the state's sanctioning of men's domestic violence were not correlated with any major transformation in legal status entailing a greater equality for women. The Constitution of 1857 made no direct reference to citizenship for women, implicitly defining citizenship as a privilege of adult men, and continued to deny women the right to vote. Women's submission to men was re-inscribed both in the Reform Laws and in the Civil and Criminal Codes. According to the Civil Code of 1870, the obligation of the husband was to feed and protect his wife and that of the wife "to obey her husband as much in domestic affairs as in the education of the children and in the administration of property."[4] A wife's capacity for legal self-representation was contingent on her husband's permission except when she was the object of a criminal case or when the case involved her husband. Legally, he had the right to administer joint property as well as any property she had brought into the marriage, though she recovered the latter in case of divorce. Divorce, a separation of bed and board, did not destroy the marriage tie and only suspended some of its obligations. Whereas female adultery was always cause for divorce (all children born to a married woman were presumed legitimate and inherited equally), male adultery could lead to a separation of bed and board only in certain circumstances, for example, if it led to mistreatment or abuse of the wife. However, though it secured their general submission, Liberal law accorded married women certain protections—"what the strong owed to the weak"—ensuring their economic maintenance, even in cases of divorce, requiring that husbands obtain their permission for the sale of property, and penalizing domestic violence. Moreover, the Civil Code of 1870 did "change the relationship of fathers and children: the age of majority was lowered from 25 to 21 and single children were freed from the *patria potestad* on reaching that age" (Arrom 1985b, 307), providing unmarried women with new freedom, as well as "granting new authority over her offspring to the widow, the legally separated mother ... and the single mother" (308).

The protection of women was part of the state's project to build a nation founded on the principles of reason, so putting an end to the chaos of the decades following Independence in which force of arms and not electoral democracy had been the main principle of succession to national leadership, a chaos that reemerged during the Wars of the Reform (1858–1861) and the subsequent French invasion of Mexico (1862–1867). The imagined geography of Liberal nationalism conflated territory, identity, and womanhood. As place of origin and source of being, the nation was represented as a Mother. However, "The People," the collective protagonist of the national, was represented as masculine as were the founders of the *Patria,* its Fathers, the heroes of the Liberals' version of history. Hence, the protection of women was linked to the consolidation and defense of the *Madre Patria* they symbolized: The feminine was a signifier not of the active subjectivity that made history but instead, of the place, of the matrix of collective being, that made history worth fighting over.

Purged of the chaos of violence, domestic patriarchy was re-inscribed in law, the embodiment of universal reason and progress. Significantly, Liberal legislation placed the most restrictions on married women as opposed to chaste single or widowed women. For the Liberals, the "conservation" and "perfection" of the species, the production of good citizens, the morality, cohesion, and order of society, and the legitimacy of public authority hinged on the stability of the heterosexual, patriarchal, nuclear family, on the wife's subordination to the husband, and on his subordination to the state—but this subordination was no longer to be guaranteed by "the abuse of force" but instead, by the rule of law.

During Porfirio Diaz's dictatorship (1876–1910), the Liberal project was undermined as state centralization took the form of a patrimonial absolutism. Much of the Liberal legislation that had rationalized authority remained on the books; indeed, the provisions made by the Porfirian 1884 Civil Code to regulate marriage, divorce, and gender identities largely reproduced those of its predecessor. However, during the Porfiriato, Liberal legislation was routinely contravened in practice, spurring resistance by members of the middle and working classes as well as the peasantry. Given the shifts in state formation during the Porfiriato, was domestic violence, even when it did not lead to serious physical injury or death, prosecuted as a criminal offense?

One approach to this question would be comparative, examining historical legal cases to understand domestic violence in Mexico "in general." I have chosen not to take this approach. First, there are few historical case studies of legal responses to wife abuse in Mexico during this period. Second, as the introduction to this collection argues, by overlooking the specificity of gender in particular places, the search for commonalities has often led to shallow insights, applicable everywhere and nowhere. Hence, I have chosen to focus on the negotiation of gender identities and relationships in domestic violence cases tried in a particular place, the *serrano* peasant *pueblo* of Namiquipa, Chihuahua, from the 1890s to 1910. However, my goal is not to use the local as a lens on the national. Instead, I integrate rather than collapse different levels of scale—national, regional, and local—and by doing so, heed the editors' call to "replace the certainty of context with the instability of place."

GENDER, PLACE, AND VIOLENCE ON
THE NORTHERN FRONTIER

Serranos or "people of the mountains" are the non-Indian inhabitants of the Sierra Madre of the state of Chihuahua, Mexico. In order to advance projects of territorial conquest and domination of Indians, particularly the Apache, the state mobilized serrano peasants for warfare in the eighteenth and nineteenth centuries. Warfare between non-Indian colonists—who called themselves "people of reason"—and Indians became construed as a struggle of "civilization" against "barbarism" (Alonso 1995). To the end of developing a "warrior spirit" among non-Indian peasant men, the state promoted a construction of gender and ethnic honor that predicated masculine reputation, access to land, and membership in a corporate community on valor and performance in warfare against "barbarians" (*bárbaros*). This state construction, deployed by serranos themselves, linked the civility and honor of men to the defense of "civilized" women. As place and point of origin, the local community was imagined as a mother. Hence, the feminine—what had to be protected—signified the stakes in the conquest of the frontier by interlinking gender, place, and community.

During this period, national, regional and local geographies of identity became interrelated and gendered in complex ways (Alonso 1995). The frontier, relative to the Center, was conceived of as a liminal zone, characterized by a greater degree of motion and fluidity, ambiguity and disorder. Planned and orderly settlement was central to the project of colonization of the frontier and to the "reduction" of its transhumant bárbaros, the Apache, meaning the concentration of dispersed populations into the units of a proper polity. In official discourse, the Apaches' perceived placelessness was a rhetorical figure that linked transhumancy to a wild masculinity, one grounded in animality, indomitability, and idleness (cf. Lyons, this volume). Located outside the social in the unbounded space of nature, this masculinity was construed as being based on force, not reason, on theft, not work, and on a liberty that was not true freedom but instead, the tyranny of the strong over the weak, of men over women (Alonso 1995, 62, 107).

The domestication of nature became key to notions of fully socialized personhood on a frontier in which the slippage between "civilization" and "barbarism" was perpetually being contained by official assertions of their irreconcilable duality. For Namiquipans, being a man required and continues to require careful management of what are considered to be the natural and learned dimensions of masculine honor. Men's "natural" qualities, valor, virility, autonomy, and mastery, are seen as the foundation for power and honor in the public and private spheres. But they are simultaneously viewed as the cause of actions that violate norms and meanings of honor and put civility and social order into question. Prey to the force of natural "passions" and "instincts," a man can "lose all sense of reason" or enter an infrasocial state where he is "out of his senses," engaging in acts of destruction and aggression that are construed as conduct that is "out of order." Hence, natural male instincts, passions, and attributes need to be tempered by "reason" and "morality" (cf. Lyons, this volume). Fully socialized masculinity entails the recognition of social obligations including the attachment to place, the ability to reciprocate, the capacity to respect others' rights to honor, the dedication to work, and the fulfillment of responsibilities to one's community and family, including the good treatment of one's wife. But the will to dominate and dishonor others and to assert autonomy can conflict with social injunctions to respect and to fulfill obligations to others and obey legitimate authorities. The everyday negotiation of the contradictions of masculinity is an arena of social contestation.

On the frontier, "civilized" women became symbols of an ethnic and sexual purity purportedly threatened by the "onslaughts of the savages." The female body and its metaphorical extension, the home, became the field of honor for women, a field whose boundaries had to be protected by the hardier sex. In Namiquipa, femininity, like masculinity, has been seen to have both natural and learned dimensions. The natural qualities of femininity are contradictory: On the one hand, like the Virgin Mary, women are seen as naturally pure, self-sacrificing, moral, and spiritual, but on the other hand, the sacralization of a virginal maternity conflicts with what are thought to be the natural facts of conception—sources of women's pollution and shame. Socialized femininity in Namiquipa is above all sexual modesty, propriety, chastity, and fidelity to heterosexual monogamy and reproductive sexuality.

Legally and divinely sanctioned marriage was the norm in Namiquipa during the Porfiriato. A multiplicity of social practices mediated the contradictory nature of femininity by imposing a symbolic closure on women's bodies and limiting their movements in space. Largely confined to the sphere of the home, women when out in public embodied the privatized character of their sexuality by wearing shawls; as one older woman described it, "their clothes dragged on the ground so that nothing could be seen except one eye." Like the Virgin in the Bible, women were subject to interdictions for 40 days after giving birth; during this period of postpartum bleeding when bodily boundaries were breached, women were confined to bed in their homes, were forbidden to go to church, were not allowed to bathe, and were subject to a number of food taboos. Since the honor of men and their ability to reproduce themselves depended on the modesty of women, women's dealings with men outside the nuclear family household were strictly monitored by male and female kin as well as community members. Married women had to contend with the jealousy of their husbands and the day to day surveillance they could resort to. Women's limited participation in agriculture was justified not only by the claim that women lacked the balls necessary for work in the fields but more importantly, because work outside the home could jeopardize a woman's honor. Until the end of the Apache Wars in 1886, outside the male protected sphere of domestic space, women were deemed to be in danger of being raped by "barbarians." They could also risk being targets of the sexual violence of "civilized" men, especially if they had contravened the norms of modesty, dishonoring themselves and their families and forfeiting rights to men's respect and protection. Such women often stayed single, and were forced into domestic service, concubinage, or prostitution in order to support themselves and their children. Gender constructions created differences among women and men as well as between them.

In Namiquipa, contradictory constructions of gender empowered women in limited ways while subjecting them to men. Over the course of more than two centuries of warfare, nonindigenous women came to be valued as reproducers of "people of reason"—in their roles as tamers of the beast in men as well as child bearers and rearers—and hence, of the crucial boundary between "civilization" and "barbarism." Women's sacred virtues have given them a key role in domestic and public religious ritual, in the "education" of children, and in the socialization of "natural" masculinity. In their capacities as mothers but also as wives, women have been seen as a central source of men's honor and virtue, of the good in men's hearts and heads. Moreover, women's work in the domestic domain and their reproductive capacities have been highly valued. Yet gender has been configured in such a way as to give a natural alibi to men's control of women's sexuality, to the division of labor, and to men's dominance in the public sphere. Lacking the physiological basis for precedence, women have been viewed as naturally incapable of wielding power since its exercise requires a hardness of heart and judgement that can only be attained by those who have balls. Though women possess "reason," their minds and hearts, like their bodies, have been seen as permeable and easily overwhelmed by softer sentiments— indexes of women's exalted nature as well as their weakness and vulnerability. Nevertheless, due to the centrality and value accorded to women's role in the home, the extent of men's authority in the domestic domain has been an arena of social ambiguity and contestation. The contradictions and ambiguities of gender ideology outlined above are highly evident in the following Namiquipan legal cases involving domestic violence.

DOWN BY THE RIVER

On September 10, 1891, Ignacio Ornelas notified the local justice of the peace that his sister-in-law, Angela Flores, had told him that her husband, José Ornelas, had hit and "abused" her.[5] Ignacio Ornelas asked the judge to investigate and to punish his brother who was quickly jailed. By defining his brother's actions as "abuse" meriting legal sanction, Ignacio was construing the use of violence to enforce domestic patriarchy as an immoral act that dishonored his family.

Angela Flores, the abused wife, described herself as 27 years old, married, and "under the respects of her husband." According to her testimony, she had found her husband talking with Soledad Frías near the river, a place outside the space of legitimate sexuality, the home, and beyond the fully socialized space of the pueblo, where others might be watching. Soledad, in the judge's paraphrase of Angela's words, "disturbs the peace of her married state with her immorality." After Angela called Soledad "an impudent abuser," her husband beat her and took her home. In her testimony, Angela Flores used the idiom of honor to construe herself as a virtuous woman whose sexuality, legitimately exercised within marriage and the space of the home, was exclusively her husband's. She characterized Soledad as a so-called public woman who belonged to all men in general and no man in particular, immorally involving herself with other women's husbands outside the space of legitimate sexuality. This public woman, Angela maintained, had affronted her honor as a wife and her right to domestic tranquility.

A 29-year-old peasant, her husband, José Ornelas, provided a different interpretation of these events. He insisted that he had no illicit relations with Soledad Frías. He admitted hitting Angela, but defended himself by impugning her morality: She had insulted both him and Soledad, saying "with insolent words, things that should not be said because they are immoral." José's version of events affirmed his moral innocence and put the blame on his wife: He merely disciplined her for her immorality, defending his honor precedence as head of the family.

Soledad Frías, who was also jailed by the judge, and who identified herself as single and 34 years old, had at least three out of wedlock children. In the preceding month, she had broken off "illicit relations" with José. She made no attempt to defend the morality of their liaison or to justify José's hitting his wife.

Angela Flores requested that her husband and his lover "be punished only in accordance with the offense committed on her person so that they recognize the duty of domestic respects." Since her husband was "the only protector whom she can count on to provide for her and her children's subsistence," she asked the judge for clemency in sentencing José. Like other women who denounced their husbands' abuses to the judge, Angela was in a contradictory position. On the one hand, she considered her husband's adultery and violence to be dishonorable and a violation of her rights to respect as a wife. Moreover, if her husband's relationship with Soledad continued, he might abandon her and their children. On the other hand, she and her offspring were wholly dependent on him economically, and a jail term would prevent her husband from giving them sustenance. Furthermore, her accusation could provoke further violence from him; but if the judge imprisoned him, she would be without his protection and, hence, vulnerable to the sexual rapacity and violence of other men. Her ambivalence is evident in her request for the judge's clemency, in her affirmation of being under her husband's respects, and in her placing much of the blame on Soledad's immorality.

On September 14, the judge sentenced José Ornelas to fifteen days in jail and prohibited him from going to Soledad's house again. In addition, the judge sought and obtained the conciliation of Angela and her husband. Though ostensibly this was a criminal case against José for "lesions," the judge sentenced Soledad to the same jail term. If the judge "protected" the limited rights of the married woman, he penalized the woman whose exercise of sexuality was outside the bounds of legitimate marriage and the space of the home and represented a threat to the stability of domestic patriarchy.

Just over a year later, Pedro Ramos, a 50-year-old married peasant, reported to the Municipal President of Namiquipa that José Ornelas and Soledad Frías had hit Angela Flores the day before.[6] Though a neighbor of the Ornelas-Flores family, Pedro bore no kinship relationship to them or to Soledad. He told the judge that he had left his house on Saturday morning and had seen Isidoro Bustillos, another neighbor, wrestling with José, trying to prevent him from hitting his wife who was exchanging insults with Soledad. Pedro rushed over, "begging them to abstain from that scandal," angering José who shouted that "nobody bossed him in what was his, that he had to proceed against his wife" and tried to attack him with a hoe. Despite José's assertion of the autonomy of his patriarchal domain and his right to beat his wife, Pedro and Isidoro continued trying to restrain him. Two female neighbors, one a relative of José's, took his wife to Isidoros's house to get her out of her husbands' way.

Angela Flores told the judge that though she was "living in good harmony" with her husband, Soledad Frías came to her neighborhood and provoked her with her presence. Her husband slapped her twice "because of that woman with whom he has illicit relations and perhaps he would have struck her again if there hadn't been people to impede him." "For her part," Angela told the judge, "she had nothing to ask against her husband but she begged the authorities, in regard to justice and her right to live well in her married state... to exile Soledad Frías from this place or to impose on her the penalty she might deserve for the scandal."

Removed from jail in order to testify, Soledad, pregnant once again, made a futile bid for feminine respectability by describing herself as married. Originally from Guerrero City, the district capital, she was now working as a domestic servant in Namiquipa. That Saturday she asked her mistress for permission to go fetch one of her daughters and when she passed by Angela's house, the latter came out "insulting and stoning me"— one Biblical punishment for women who had committed sexual transgressions—with no provocation since she had not had "relations in private life" with José for a year and three months. José, who was jailed also, denied hitting his wife or having sexual relations with Soledad. By his account, Soledad grabbed his wife in order to stop her from stoning her, and he and Pedro Ramos separated them.

In contrast to the judge in the prior case, this judge declared that no crime had been committed, only violations of the Edict of Good Government. He did sentence José to eight days of reclusion but for having broken out of jail to "protect" Soledad, not for beating his wife. As in the prior case, Soledad was penalized for her shamelessness, receiving the same sentence as José, for having gone near Angela's house; note that she was punished for her anomalous presence near the place where legitimate sexuality should be exercised, the conjugal home, the domain of the private not the public woman. Both were admonished not to repeat the "abuses" that

Angela accused them of committing. The judge concluded that his decision was based on the power the law gave him to "impede scandals and proceedings that perturb public tranquility and morality." What he sanctioned verbally was José's beating of his wife in public, implying that a man did have a right to do this in private; in this instance, the relationship between gendered violence, morality, and space, and the social relations and obligations thus coded, was crucial to the verdict. As in other cases of domestic violence elsewhere, publicity offered protection to abused wives.[7]

The cases I have presented have a number of points in common with others in the Namiquipan judicial archives. Women who were hit by men were generally married to them; in most cases, blows were administered with the hand and resulted in bruises rather than in more severe physical injuries.[8] Though a few women specified that their husbands had hit them before, most did not state this. Husbands hit their wives to enforce their precedence in the domestic domain in the face of what they perceived as their wives' challenges to their authority and to assert their rights to: (1) commit adultery; (2) have exclusive access to their wives' sexuality and generativity; (3) have complete control over household economic resources and their wives' property; (4) command their wives' domestic services whenever and however they wanted; and (5) use violence to discipline their children, particularly sons. The first two appear to be the most common motives for domestic violence cited by both men and women, though since the judicial archive is incomplete, surviving cases may not be fully representative of all the cases tried.

Overall, accused husbands either denied hitting their wives or justified their use of violence, particularly in the "privacy" of their homes, as morally and naturally legitimate discipline or as legitimate defense of their honor when they suspected their wives were betraying them with other men. But some men's attitudes were more ambiguous. There were those who blamed their violence on drunkenness, thought to be a state marked by a temporary loss of reason and, hence, self-control and intentionality, as well as a regression into natural masculinity. Invoking inebriation as an alibi could lead to a reduced sentence; however, it also implicitly denied the rationality and morality of wife beating. Moreover, the pledges made by some abusers to their wives betray some ambivalence about the morality of domestic violence and male adultery; for example, one husband promised his wife that "in the future he would treat her with the decorum that a wife deserves, offering to never again put a foot in the house of his concubine."[9] Significantly, in most of the cases where the couple was conciliated by the judge, husband and wife stayed together and no further legal complaints about domestic violence were presented.

Simply by denouncing their husbands' violence to the judge, the municipal president, or a male or more rarely, female relative, who then reported it, women were contesting the moral legitimacy of using violence to enforce patriarchy and affirming their right to "good treatment" and respect, which included their husbands' faithfulness, courtesy, trust, and economic maintenance. Many women characterized their husbands' violence as *"un ultraje,"* an outrage, offense, or insult, a term also used as a gloss for rape, implying that domestic violence affronted their honor and dishonored the abuser. The rhetoric of honor sacralizes persons and the ideal sphere surrounding them: For women, the forcible breaching of this sphere was both dishonoring and dishonorable. Some women affirmed that their husbands needed to learn "the duty of domestic respects." Others characterized their husbands' violence as "bad treatment," as "mortification," or as "an atrocious and ferocious trampling of his family under foot." One woman complained that her husband had "inflicted cowardly and perfidiously the offense of outraging or insulting her by word and deed,"[10] while another maintained that her spouse's violence was an indication that he was "out of his senses."[11] Importantly, these women's definitions of the identities "husband" and "wife" and of the extent of men's domestic authority and responsibilities conflicted with those of their husbands'. Women questioned the morality of their husbands' adultery. They contested the way their own property as well as joint property was administered by their husbands. One woman, for example, claimed her husband beat her because she sold two of her cows.[12] Another included in the motives for her husbands' abuse her refusal to give him permission for the sale of some land unless he paid her the value of the property she had brought into the marriage because she was worried her family would become wholly destitute due to his "dissipated conduct."[13] Women also questioned husbands' rights to beat them due to jealousy, and the implicit challenge men's policing of their sexuality posed to their own capacity to abide by the norms and values of modesty. Such contestation of the link between masculinity and authority suggests that significations of the home were unstable. Men and women might have agreed that as the locus of domesticity, the home was a feminine space, and that as the domain of patriarchal authority, it was a masculine space. Yet legal cases suggest that there were important differences in men's and women's definitions of appropriate domesticity and responsible authority.

Men's violence pitted women against each other. If a wife felt that the immediate cause of the blows was her challenging the morality of her husband's adultery, she would frequently lay at least equal, if not more, blame on the other woman. Indeed, the few cases of violence between women in the judicial archives involve wives and concubines. By "publicly identifying another woman as the cause of her marital problems, a wife might indirectly pressure her husband to fulfill his patriarchal responsibilities" (Chambers 1999, 33). But wife abuse also provided bases for solidarity among women. In some cases, mothers and sisters of the beaten wives supported them against their husbands. In other cases, women neighbors or relatives of the abuser assisted the victim.

In order to successfully contest their husbands' ill-treatment in the realm of law, women had to establish their own honor and morality; in effect, they had to affirm their general submission to domestic patriarchy even if they contested some of its particular manifestations. In contrast to the case of Pangor, Ecuador, studied by Lyons, this entailed embodying femininity. In addition, women's denunciation of abuse was tempered by their dependence on men, and by the risk that too strong a censoring of their husbands' actions would either fuel their men's anger and violence or lead to abandonment. In some cases, such as the ones presented earlier, the woman asked the judge for clemency in his sentencing of the husband, or explicitly pardoned the man for the sake of her children, since his confinement in jail or fining represented an economic hardship for the family. In these cases, the judge conciliated the couple. However, in other cases, women affirmed that they could no longer bear the man's "ill-treatment" and asked for a divorce as well as custody of the children. Though some of these women went to live with members of their family of origin, others remained as heads of their own families. I have not found a single case in which the woman was forcibly returned to the conjugal home, even when her husband demanded it. However, as a rule, beaten wives were not contesting domestic patriarchy as such but, instead, men's failure to temper their tendencies to sexual rapacity, violence, and mastery with reason and the norms and values of honor as virtue.

I have discussed domestic violence and some of these legal cases with women friends in Namiquipa. They were fascinated by the cases, comparing them to soap operas. Like soap operas, the law serves as an incitement to discourse about gender, sexuality, and other privatized aspects of personhood and practice that are simultaneously defined as matters of public, and hence, state, interest. My friends share their predecessors' views on the immorality of husbands' use of violence against wives. They represent a man's beating his wife as a cowardly abuse of force: A husband should use his strength to protect and not to offend his wife, they maintain. Some women have told me that wife-beaters are *machos*—that is, they possess the natural qualities of masculinity that are the basis for precedence—but they are not *hombres,* that is, they lack honor and virtue and are not fully socialized men. Such men's actions reflect badly on the honor of their families of origin and imperil the stability of their families of procreation, setting a "bad example" for the children and leading to conflicts between fathers and sons, and sometimes daughters, who try to protect their mothers.

Though many of the men accused in domestic violence cases affirmed their right to "correct" their wives, not all men shared this view. Given that men's honor depended on that of their close women kin, it is not surprising that fathers and brothers of abused wives would formally accuse husbands. Though one could argue that abused women's male kin were defending their own status rather than questioning a husband's right to discipline his wife, the support of men not related to abused wives suggests otherwise. Men who were neighbors or kin of either party often risked the wrath of the abuser by intervening to stop the beating. Accusations of wife beating were brought before the judge by men, such as Ignacio Ornelas, who were relatives of the abuser—fathers or brothers—or by male neighbors, such as Pedro Ramos. In addition, local male judges, who were community members, as opposed to legal professionals, with a great deal of leeway in the interpretation of laws, generally censured wife abuse.

Most of the legal complaints of domestic violence were followed up and the accused was arrested as soon as possible and kept in jail until a verdict had been reached. A guilty verdict was reached in most of these criminal cases, and even when wives pardoned their husbands or asked judges for clemency, husbands were generally sentenced to between 8 and 30 days in the local jail. In cases where male adultery was an issue, judges frequently prohibited the husband from seeing his concubine again. Some judges censured wife beating in very strong terms. For example, one judge condemned "the scandal with which this individual is accustomed to outrage or insult his poor family in a fairly cowardly and vile manner."[14] In another case, a wealthy merchant's defender argued that the woman had brought the violence upon herself and "that he only did it to maintain the legitimate right that he who commands has to repress in the home and in private the faults of his wife." The judge responded to this by saying that the husband was not "authorized to offend his wife for any reason with legal right"[15] and sentenced the merchant to 15 days in jail.

Overall, local judges promoted the stability of domestic patriarchy by serving as arbiters in marital conflict, using the clout their position afforded to conciliate the couple and encourage the good functioning of the family. Judges used their power to support women who abided by the norms and values of honorable femininity. Like the state's laws, the judges' verdicts did not represent a critique of women's subordination to men, but instead, support for a form of patriarchy based more on "civilized" reason than "barbaric" force.

The state, as the editors of this volume point out, "is not only present in its formal institutions, but is also incarnated in the negotiations that take place locally" (11). Namiquipan men routinely ignored or challenged state laws, injunctions, and policies they did not consider legitimate. Legislation regarding wife beating was applied in Namiquipa only because it resonated with some dimensions of local ideology and gender constructions. Men's treatment of women was one of the indexes of possession of reason, the key mark of civility and privileged ethnic status in Namiquipa. From the colonists' point of view, Apache men, who lacked reason and were the epitome of unruly natural masculine instincts and passions, mistreated their women, making them do all the work and correcting faults like adultery with "barbaric" punishments such as cutting off the nose of the woman. Moreover, as the policies of the Porfirian regime increasingly jeopardized peasants' claims to land, Namiquipans' assertion of civility played a key role in their attempt to retain corporate agrarian rights. In addition, during the Porfiriato, effective suffrage became increasingly ineffective as elected leaders were replaced by *caciques*—local strongmen who were clients of regional patrimonial leaders. Numerous Namiquipan petitions construed *caciquismo* as an abuse of force that contravened the will of the people and the injunctions of honor as virtue: If the Liberals had made a connection between the construction of authority in the public and private spheres, so too had Namiquipans (see Alonso 1995). Significantly, one woman used the idiom of the struggle against caciquismo to characterize her husband's abusive behavior, accusing him of "despotism and tyranny."[16]

Thus the boundaries between reasonable leadership and abuse of force, between honor as virtue and honor as precedence, between private and public spheres, between protection and discipline, between obedience and defiance were contested: Constructing themselves as honorable women deserving masculine protection, abused Namiquipan wives paradoxically challenged the extent of their husbands' power while affirming their own submission and reproducing hegemonic constructions of femininity. By contrast, abusive husbands construed their wives' actions as disobedience meriting legitimate discipline. Surviving cases indicate that the law was applied more often than not, and that many men and women in the community censured domestic violence as an abuse of force and an offense to community and family honor and morality. Overall, for Namiquipan women, Liberal legislation was an important resource in their attempts to negotiate a limited empowerment within the terms of their overall subordination. Such spaces of negotiation are revealed when we conduct our struggles as well as our analyses on two interconnected levels, "an ideological, discursive level which addresses questions of representation ... and a material, experiential, daily-life level which focuses on the micropolitics of work, home, family, sexuality, etc." (Mohanty 1991, 21).

CONCLUSION

This chapter has outlined multiple intersections of gender, space, and place. If places are "best understood by thinking about particular sets of intersecting social relations" (Cravey, this volume, 286), the converse also holds true: Social relations are best understood by thinking about particular sets of intersecting places and spaces. The well ordered polity attempts to fix subjectivities by inscribing them in the topography of day-to-day life and to localize the practices of subjects, facilitating their regulation while legitimating protest only when it is situated in particular political spaces. Power has a geography. The meanings and consequences of violent acts vary according to their placement in space and the social locations of those involved. Spatial boundaries, such as that between the public and the private, are often icons and indexes of social boundaries and, hence, are points of regulation and contestation.

My research depicts the complex interplay between constructions of gender and authority in the public and domestic spheres. Policies that favor women in limited ways—such as nineteenth-century Liberal legislation— can be part and parcel of new forms of subordination to men. In twentieth-century Mexico, as Varley (2000) demonstrates, judicial redefinition of the marital home modernized patriarchy, altering gendered configurations of power within the household without eliminating women's overall subordination to men. Although Foucault writes that "power in modern societies has not in fact governed sexuality through law" (1980, 90), he accepts the received notion of the juridical as functioning mainly through prohibition too readily and, since he argues that modern forms of power are primarily productive rather than repressive, he underestimates the importance of law. I have argued that it is precisely this opposition between the productive and prohibitory character of power that needs to be undermined because the prohibitions of law as code and as practice are productive of subjectivities, of normative forms of classification, of social relations, of forms of power/knowledge. Law, as one of the key points of condensation of power in modern states, has played a crucial role in the construction of gender, sexuality, and patriarchy.

In writing about domestic violence in Mexico from my location on this side of the border, I have tried to both undermine Anglo stereotypes of Hispanics while criticizing patriarchy in Mexico. International institutions such as USAID and the World Bank, political institutions and the media in "the North," and even mainstream feminism frequently define the subordination of women of color in terms of an opposition between tradition and modernity. Whereas the narrative of modernization that underpins these images makes the North the site of reason and the South the site of an irrational male violence, female passivity, and silence, I have shown how rationality and rationalization have been integral to state formation in Mexico, providing alibis for women's subordination. Furthermore, I have shown how, in one Mexican rural community, the use of violence to enforce domestic patriarchy was contested by both women and men.

NOTES

Field and archival research was funded by the Social Science Research Council and the Inter-American Foundation (1983–1985), the Center for Latin-American Studies, University of Chicago (summer 1986), and a Mellon Summer Grant by the Institute for Latin-American Studies, University of Texas, Austin (summer 1989). I have drawn freely from my book (Alonso 1995) in this chapter. An earlier version of this chapter, "Rationalizing Patriarchy: Gender, Domestic Violence, and Law in Mexico," was published in *Identities* 2(1–2): 29–48, 1995. See this version for an extended discussion of the legal history. I thank the editors and anonymous reviewer of this volume for their helpful comments and suggestions for revision.

1. Article 39, "*Constitución Política de la República Mexicana*," February 12, 1857, reproduced in *Código de la Reforma o colección de leyes*.
2. Articles 501–509, Articles 511–539, Article 554, *Código penal para el Distrito Federal y la Baja California*, December 7, 1871, reproduced in Dublan and Lozano 1879, vol. XI, 597 ff. The "first Mexican Criminal Code" of 1871, the first Mexican Civil Code of 1870, and the revised Civil Code of 1884, were not federal in scope, governing only the Federal District and those territories that had not attained the constitutional status of States in the Federation. However, the Federal District codes served as models for those adopted by the States, often with little or no alteration.
3. *Codigo de reforma* no. 52, July 23, 1859, in *Codigo de la Reforma o Colección de Leyes*, all quotes are my translations.
4. Article 201, *Código civil para el Distrito Federal y la Baja California*, December 13, 1870, reproduced in Dublan and Lozano, vol. XI, 201 ff.
5. *Archivo Judicial del Municipio de Namiquipa* (AJMN), box 2, September 10, 1891, criminal case against José Ornelas for lesions, all quotes are my translations.
6. AJMN, October 24, 1892, criminal case against José Ornelas and Soledad Frías for lesions; all quotes are my translations.
7. Article 510, 1871 *Código Penal*, specified that "blows [*golpes*]" that did not cause lesions could only be prosecuted at the request of the aggrieved party or if they were delivered in a public place.

8. I have found no cases in which women were permanently injured or killed. However, a few women stated that their husbands threatened to kill them and in one instance a drunken husband shot at his wife and missed.

9. AJMN, Isabel Iturrales against Trinindad Flores for adultery with her husband Blas Camarena, date unclear.

10. AJMN, June 8, 1897, conciliation filed by María de la Luz Dominguez de Rivera.

11. AJMN, Box 3, April 19, 1897, criminal case against Margarito Ramirez for hitting his wife, Isidora Burciaga.

12. AJMN, Box 3, April 19, 1897, criminal case against Margarito Ramirez for hitting his wife, Isidora Burciaga.

13. AJMN, June 8, 1897, conciliation filed by María de la Luz Dominguez de Rivera.

14. AJMN, Box 1, April 7, 1890, case involving María Luisa Ivarra de Ivarra, forwarded by the *Juez Rural* of the *Hacienda de Providencia* to the Namiquipan Justice of the Peace.

15. AJMN, July 3, 1904, criminal case against Jesus M. Gutierrez for lesions.

16. AJMN, box 6, May 15, 1909, Pedro García demands that his wife return home or that she be subjected to judicial deposit; my translation.

REFERENCES

Alonso, Ana María. 1995. *Thread of Blood: Colonialism, Revolution and Gender on Mexico's Northern Frontier.* Tucson: University of Arizona Press.

Arrom, S. M. 1976. *La mujer mexicana ante el divorcio Eclesiastico (1800–1857).* Mexico: SepSetentas.

———. 1985a. *The Women of Mexico City, 1790–1857.* Stanford: Stanford University Press.

———. 1985b. "Changes in Mexican Family Law in the Nineteenth Century: The Civil Codes of 1870 and 1884." *Journal of Family History* 10(3): 305–317.

Chambers, S. 1999. " 'To the Company of a Man like my Husband, No Law Can Compel Me': The Limits of Sanctions against Wife Beating in Arequipa, Peru, 1780–1850." *Journal of Women's History* 11(1): 31–55.

Codigo de la Reforma o Colección de Leyes, Decretos y Supremas Ordenes Expedidas desde 1856 Hasta 1861. 1861. Mexico: Imprenta Literaria.

Collier, Jane, B. Maurer, and L. Suárez-Navas. 1995. "Sanctioned identities: Legal Constructions of Modern Personhood." *Identities* 2(1–2): 1–27.

Dublan, M. and J. M. Lozano. 1879. *Legislación mexicana o colección completa de las disposiciones legislativas expedidas desde la Independencia de la Republica.* Mexico: Imprenta del Comercio.

Foucault, Michel. 1980. *The History of Sexuality, Vol. I: An Introduction,* translated by R. Hurley. New York: Vintage Books.

Mohanty, Chandra T. 1991. "Introduction: Cartographies of Struggle: Third World Women and the Politics of Feminism." In *Third World Women and the Politics of*

Feminism, edited by Chandra T. Mohanty, Anne Russo, and L. Torres. Bloomington: Indiana University Press, 1–47.

Varley, A. 2000. "Women and the Home in Mexican Family Law." In *The Hidden Histories of Gender and State in Latin America,* edited by E. Dore and Maxine Molyneux. Chapel Hill, NC: Duke University Press.

6. Placing Gender and Ethnicity on the Bodies of Indigenous Women and in the Work of Bolivian Intellectuals ∽

Susan Paulson

Policies and programs established in Bolivia in the 1990s expressed a new vision of political agency and citizenship: a multicultural, pluri-ethnic, and gender-sensitive vision that broke the longstanding assimilationist paradigm and promised greater respect for diverse identities and lifeways. This paper explores practices and discourses of dissimilar social groups, and asks how their dynamic coexistence has contributed to unique expressions of identity and alterity articulated in Bolivia. My discussion draws on detailed ethnographic research in three sites (a government ministry, a university research project, and the everyday lives of indigenous women and men), carried out during the decade that I lived in Bolivia.

Rather than address identity as something that is generated by and rooted in specific geographic places and demographic groups, I explore the intense struggle over identity that goes on between apparently disparate places and peoples. My purpose in bringing these sites together is to grasp the tense interplay among local, national and global forces, ideas, and actions that give shape and meaning to cultural and gender difference. My object of study here is not a noun—a gender group, an ethnic community, or a specific legislation—but rather a verb, some kind of process through which such things as gender identities and ethnic communities are built, named, negotiated, and contested.

Numerous factors set the scene for the emergence of remarkable new visions of identity and citizenship in Bolivia. Shortly after the 1993 elections, the Sanchez de Lozada administration established a Ministry of

Human Development with a Secretariat of Ethnic, Gender, and Generational Affairs. Some Bolivian intellectuals, well aware that politics often masquerade as culture, were suspicious of top-down efforts to forge more equitable forms of social participation. At the same time, they were thrilled with the secretariat's unprecedented support for ethnographic research, community-based projects, publication of scholarly writing, and ongoing forums for national debate about identity issues.

The challenge for engaged scholars who were invited to join government agencies was to put their progressive ideas to work in institutional landscapes deeply engraved by hierarchical relations and dominated by neoliberal ideologies. Projects throughout Latin America have tended to place gender on the bodies and lives of women, and locate ethnicity in indigenous groups, thus conveniently sidestepping the need to question or transform broader social structures and relations. In attempts to move beyond this narrow compartmentalization, key members of the secretariat promoted a fascinating new spatial trope called *transversalization,* which implies extending considerations of gender and ethnic issues into a wide spectrum of policies and projects. Words of the first secretary of Ethnic, Gender, and Generational Affairs, anthropologist Ramiro Molina, demonstrate how these progressive new strategies were still couched in longstanding hierarchical frameworks. "My idea was to integrate ethnic awareness *horizontally,* among ourselves, government functionaries placed throughout the administration. We put people in the Ministries of Education, Popular Participation, and Rural Development in order to *transversalize* those ideas. At the same time, we develop a *vertical* axis of work with indigenous organizations. How can we advance policies for and with differentiated groups if there are no paths for them to reach us?" (Interview with Ramiro Molina, La Paz, March 17, 1998) [italics mine].

ETHNOGRAPHIC SCENARIOS

The following three scenarios reveal glimpses of struggles to shape representations of gender and ethnicity played out during the mid-1990s in the Department of Cochabamba, in a national theater featuring identity politics designed to recognize and respect subaltern social groups alongside neoliberal economic politics that threatened the material well being of these same groups.

"Bolivia, free, independent, sovereign, multi-ethnic and pluri-cultural, constituted as a united republic, adopts for its government a form of representative democracy grounded in the union and solidarity of all Bolivians"

So begins Bolivia's 1993 revised constitution, which shares with widely acclaimed Popular Participation legislation from 1994 the goals of recognizing diverse Bolivians and fostering greater local involvement in political processes. After decades of citizen-building projects aimed at incorporating diverse actors into elite European models of citizenship, these reforms represent a dramatic about-face in which cultural diversity and gender identity take center stage in national politics (Healy and Paulson 2000; Luykx 1998). The impressive cadre of intellectuals recruited for this new government project included many whose lives had been shaped by struggles *against* repressive governments, struggles often designated as for and about groups who suffered class, ethnic, or gender oppression.

The first director of the Subsecretariat for Gender Affairs for the Department of Cochabamba, architect María Isabel Caero, with whom I collaborated on several initiatives,[1] addresses the challenge of moving her opposition politics into government offices. "Being part of the Subsecretariat of Gender signifies a repositioning from earlier times when we found ourselves contesting that which we considered the crudest expression of domination, the State, which we characterized as capitalist, authoritarian, and patriarchal. Rethinking the State in relation to women means accepting changes occurring in the world. It means accepting that neoliberal measures do exist, but that coexisting with them are desires to develop innovative proposals fostering political, conceptual, and methodological advances toward improving the condition and position of women in our country" (1997, 71).

At the same time that they rethought the state, members of the subsecretariat tried to break free of stereotypical "women only" approaches to gender, "We had frequently dedicated ourselves to training, educating, implementing productive projects with women, supposing that in this way we could improve their situation. We dreamed that for us, and for them, consciousness of subordination was sufficient to overturn it. Nevertheless, in time we realized that if we do not transversalize gender into the macro policies of this country, any improvement would be slow and difficult" (Caero 1997, 72). Although many of those involved understood gender as a cultural system that *transverses* all actors and relations in Bolivia, the bulk of studies and projects actually carried out by the Subsecretariat still ended up locating gender in the lives and issues of women. Among the factors that reinforced this pattern (discussed in Paulson and Calla 2000) was the unit's structural position within the social services branch of departmental government. Thus placed, gender staff had to resist becoming a ghetto of women administering women's welfare, "The Departmental Unit of Gender Affairs *should* introduce gender concerns into public policies in all instances of the

138   *Susan Paulson*

Prefect and all Municipalities of the Department, it *should* form part of the Secretariat of Human Development. Its current place in the structure of the Prefect biases its orientation towards traditional social assistance for women, prioritizing only their roles as wife-mothers" (Caero 1997, 72, italics mine).

While gender advocates shared with those who worked on ethnicity the struggle to transversalize their concerns across broad political spaces, the gender camp alone pushed into so-called private places. As Caero testifies, "I talk of radical democracy because in these times in which democracy is considered the best form of governance, I embrace democracy not only in the public sphere, but also in the most intimate human relations" (1997, 71). In light of this willingness to radically rethink political practices, Caero and her colleagues were bitterly surprised to find among their harshest critics some men who had been militants with them in the struggle for democracy. These leftist men located the roots of feminism in imperialist nations, and resisted the introduction of gender-focused programs in Bolivia as "impositions of the World Bank and the International Development Bank" (1997, 77). Similarly, some champions of ethnicity saw the innovative position of the Gender Unit as a threat to indigenous autonomy, and vehemently attacked what they perceived as feminist meddling in the private lives of Bolivia's ethnic groups.

"Andean women do not exist individually, outside of the family and the community, women do not think, need, or act outside of the couple"

Political conflicts between advocates of gender and of ethnicity have also contributed to a polarization of these categories in the realm of scholarly research. A Dutch program that supported teaching and research in Bolivian universities with the objective of better understanding irrigated agriculture from a gender perspective became one of many battle scenes. As a colleague who knew the research project, but was not a member of either of the teams competing for future funding, I was asked to serve as moderator and discussant at a research seminar held in 1995. I summarize the proceedings here on the basis of my notes and documents written by participants.

A group led by university professors María Esther Pozo and Christiane Tuijtelaars had received a major grant to carry out research, the results of which were published in a book, two bachelors theses, and several academic articles (Pozo 1994; Tuijtelaars et al. 1994). Their approach, which I call "classic gender analysis," used concepts and methods promoted by first world scholars like Moffat et al. (1991) to measure individual women's access to and control of productive resources and political decisions. The study's unprecedented efforts to apply this gender analysis model to detailed productive processes in Bolivia raised many critical questions, among them doubts about the cultural appropriateness of the research model and of the concept of gender in which it is grounded.

A second group of scholars who identified themselves as Andeanists presented an alternative proposal for the next round of research grants. Some members of this group had engaged in political activism with members of the first group in the 1970s and 1980s, and two had since gotten involved with the Andean fundamentalist movement Proyecto Andino de Tecnologías Campesinas (PRATEC) (see Apffel-Marglin 1998). In a forceful critique of the classic gender analysis, this Andeanist group argued that in order to understand irrigation and farming systems in Andean communities it is necessary to study qualitative and symbolic aspects of life that are culture specific and not addressed by ethnocentric models developed by Northern feminists.

When the classic gender group presented statistical information and opinions gathered through interviews with individual women, a member of the Andeanist group retorted, "Andean women do not exist individually, outside of the family and the community, women do not think, need, or act outside of the couple. We must focus our research on community, rejecting individualism as a destructive Western concept." When the first group decried the exclusion of women from farmers' syndicates, producers' cooperatives, irrigation associations, and other formal institutions, Andeanists emphasized the need to study women's and men's participation in rituals, work parties (*mink'a*), exchange networks (*ayni*), and other non-Western, noncorporate forms of organization and action. When the gender researchers calculated wealth, property, labor, offspring, and residence based on nuclear family units, their opponents insisted that extensive kin and ritual networks, together with traditional polities called *ayllus,* are the only valid units of analysis for understanding human and natural resource management in the Andes. Finally, when the gender group presented sex-disaggregated data demonstrating unequal land tenure, water rights, and livestock ownership, a member of the Andeanist group insisted that "those data refer only to paper documents produced to show to officials in order to obtain loans and technical support. They are meaningless among the Andean community where what really matters is the personal and spiritual relationships that men and women maintain with the land (*Pachamama*), water (*Q'ocha*), animal beings, and other natural forces."

Seminar participants construed the gender-sensitive approach and the ethnic-sensitive approach as methodological opposites and ideological enemies. Yet both groups coincided in associating ethnicity with tradition and placing the idea of gender within modernity. Ethnic traditions, for the first group, are barriers to be overcome through better education and improved standards of living; for the second, such traditions are unique and untouchable lifeways to be glorified, preserved and protected. The first group understood gender analysis as a positive tool for democratizing and modernizing local identities and relationships that were traditionally discriminatory. Members of the second group saw gender analysis as an imperialist tool for annihilating the complementary and harmonious identities and relationships that characterize traditional Andean communities, following ideas published by PRATEC co-founder Eduardo Grillo. "Imperialism has chosen the gender and development focus with the specific goal of damaging the family and the women in our countries, because they are the fundamental nucleus through which the great diversity of indigenous cultures of the world are regenerated" (1994, 15).

"When a woman is hoeing potatoes in her field she is a campesina, but when she goes to the city to sell her potatoes she is a chola"

With these words, Faustina Fernández answered my awkward questions about identity categories like *campesina* (rural peasant woman) and *chola* (merchant or urban indigenous woman); her practice showed me serious limitations in the static ways in which I was trying to understand identity. Early one morning, Faustina pulled a ratty sweater over her "Murphy Crane and Erection Company" T-shirt and tied on a dark *pollera* skirt reaching almost to her ankles, before waking her daughters to help prepare breakfast for the group of children, siblings, nephews, and others who would gather to harvest her potato field.

Throughout the long day of cooking, serving, digging, harvesting and sorting potatoes, Faustina performed a certain gender and ethnic identity in the ways in which she administered and implemented each task in coordination with others from her local community. In the late afternoon, however, her identity shifted as she entered into transport negotiations with a mestizo trucker. Leaving her sister in charge of the final sorting and bagging, Faustina hurried back to her patio where she combed her hair, rebraiding it with shiny hair pieces and brightly colored tassels. Changing into her best market clothes, a short pink pollera skirt and tight lace blouse glittering with plastic pearls, she explained, "the *transportista* is due at six o'clock to load the potatoes, and if he thinks I am some dirty Indian he'll cheat me in the portion of potatoes he takes in exchange for transporting my cargo."

After cutting the deal and getting her potatoes loaded, Faustina said goodbye to her family, assigning chores to each child. She spent the night in the back of the truck, bouncing along on bags of potatoes together with others who were traveling the long road to Cochabamba. Arriving at the market before dawn, Faustina arranged her produce in a market stall rented by a cousin who lived in the city. She was careful not to intrude on the space of the neighboring chola vendors, as they resented her presence, calling her a clumsy campesina. Nevertheless, Faustina competed successfully for the attention of passing customers and skillfully negotiated identities in her interactions with them. She played up her hometown identity as Mizque potatoes are known in the city for their quality. She conversed and joked merrily with male customers in Spanish, aware that among urban mestizos flirtatious relationships—as well as sexual encounters—with indigenous women enhance manly identity. With urban housewives, however, Faustina ingratiated herself by responding to their address of "*waway*" (my child) with humble poses and phrases sprinkled with her native language, Quechua. She knew that by emphasizing her Indian ethnicity she could better please clients whose own sense of identity (whiteness, educatedness, cleanliness, female purity) depends on their sense of superiority to her.

Whereas Faustina Fernández's articulations and practices of relational identity encourage more imaginative thinking about the relationship between gender and ethnicity, possibly influencing visions that Caero and her colleagues hoped to implement in new state projects, the kinds of bitter antagonisms expressed in critiques of the Gender Unit and in debates in the research seminar function to limit the radical potential of the concept of transversalization. How is it that innovative activists and scholars who share a history of political struggle and a commitment to emancipatory change have gotten so mired in divisive polemics surrounding gender and ethnicity?

US AND THEM

The large proportion of Bolivians who self-identify as indigenous and the high level of contact among people of different cultural orientations contribute to intense discussions about ethnicity throughout the country. At the same time, the separation of hierarchized social spheres, together with the concentration of political power in elite European-identified society, have allowed dominant national discourse to inscribe concerns about ethnicity on the bodies of indigenous-identified others. Indeed, throughout Latin America, scholarly discourse and political programs have tended to direct both ethnic and gender attention to marked groups of so-called marginal others, often epitomized by people like Faustina Fernández, discussed above. This essentialist focus draws analytic attention away from privileged social groups like the researchers and the government employees who populate the other two scenarios and conveniently allows those in privileged positions to avoid addressing the power relations that differentiate individuals and groups.

How has the national and international positioning of social scientists framed analytic categories of gender and ethnicity and shaped the possibility of representing either marginal or powerful others? Part of the answer lies in the analytic framework dominant amongst international organizations that not only influence, but actually finance, the government bureaucracies and university research projects in question, as well as many other intellectual and social forums in Latin America. Warren and Bourque identify there a "language of difference [that] portrays other societies as victims (as passive 'targets' of programs or 'receptors' of technology) rather than as constructors of their own cultural understandings of change and technology" (1991, 302). Many Bolivian scholars and government professionals depend on support from international organizations in arrangements that tend to place them in the role of (global) agents, working to help or benefit the (local) masses in their own country. These opportunities taste of bitter irony for actors like professor María Esther Pozo and subsecretariat director María Isabel Caero, who once sang "A Desalambrar" (see introduction to this volume) in youthful zeal to overcome the class and cultural barriers that divided their country.

Policies and projects are usually designed to recognize and help poor people, indigenous people, women, single mothers, homosexuals, illiterates, and the like. What is ironic in the Bolivian case is that these so-called marginal groups make up a vast majority of the national population. In keeping with dominant global discourses, Bolivian newspapers and television announcers diligently repeat key markers of otherness that construe Bolivian majority groups as inferior marginals. We frequently heard about Victor Hugo Cárdenas, "the indigenous vice president," while no one referred to Goni Sanchez de Lozada as "the white president." One of the few women legislators in Bolivia, María Lourdes Zabala, has been represented in the press as "*la diputada mujer,*" or "*la diputada feminista.*" Another female parliamentarian, Remedios Loza Alvarado, who wears ethnically marked clothing and was a presidential candidate in 1997, isn't even given a last name by the media: she is simply dubbed, "*la Chola Remedios*" or "*Comadre Remedios.*" At the same time, however, no self-respecting journalist would announce that "the European-male Blattman" voted thus on a given legislation, or refer to Tito Hoz de Vila as "the pro-masculine minister of education." These white men simply have names and political titles; their lack of ethnic and gender markers, crossed by a privileged class position, leads us (and them) to believe that they have nothing to do with gender or ethnicity, let alone poverty. Paradoxically, these same national elites are marked as poor, marginalized others in international geopolitical discourses (see Coronil 1992).

Media and politicians frequently situate subaltern others in a different place from dominant groups, a place that can then be studied and improved from the outside. Gupta and Ferguson attempt to unmask this deception by stressing that place-making always involves a construction, rather than merely a discovery, of difference, that identity is not a thing that grows out of bound places, but rather is a mobile, often unstable *relation of difference* (1997, 13). And, while the intensity of that relation pushes Bolivians in all sectors to debate about difference, the tendency to locate differences in socially distant places (rather than in relations between and within places and people) remains powerful.

Faustina Fernández knows well how to play "the other," and she does so as stereotypically feminine and indigenous as can be. Yet as I watch her deftly transform herself, I cannot make any kind of labels stick. Her own discourse, stressing the *when* and the *where* of identity, reveals that gender and ethnicity are not simply contained within her body, but are actively generated in the contexts and relations in which she engages. And these contexts and relationships are shaped by the vital presence of unmarked groups, ranging from superior mestiza housewives to lecherous male clients.

Gender and ethnic analyses that transverse society must move beyond focus on women like Faustina Fernández to encompass the politics of knowledge and identity at play in places such as universities and government agencies. However such a move would require scholars and professionals (including myself) to abandon the comfortable us–them distinctions on which our work rests, and to recognize the historical positions, privileges and powers that facilitate our ongoing efforts to analyze—and to shape—identity discourses.

SOCIAL POSITIONING AND REPRESENTATION

Consideration of some historical positions and biographical experiences of Bolivia's middle/upper class (a class that engenders both intellectuals and government professionals), together with factors that drew some of these elites into movements to politicize women's and indigenous issues, facilitates the dismantling of the us–them dichotomy.

Beginning in the 1940s, intellectual leaders inspired by nationalist ideologies and Marxist/Leninist visions established specific power/knowledge relationships with poor and exploited groups, most notably miners, and later peasants. Vanguardism became a powerful mode of representation and decision making for educated elites who championed the causes of oppressed others. While the paternalistic impulses channeled through these movements are still manifest today in efforts to "help" and "empower" marginalized people, these are increasingly defined in ethnic and gender, rather than class, terms.

Many Bolivians who are currently involved in ethnic and gender politics came of age as leftist militants who risked their lives fighting for democracy and political–economic justice during the 1970s and 1980s. Some later formed nongovernmental organizations, which became a major channel for expressing political ideas and actions and which often construed efforts to provide alternative health, welfare, or education services to marginal sectors of the population as forms of opposition to military dictatorships and state capitalism (Gill 1997). In more recent turns of fate, ex-militants like María Isabel Caero and Ramiro Molina moved into the government itself to head programs addressing gender and ethnic inequalities. Caero recognized the ongoing influence of their political heritage, "For me it is important to talk about ideological commitment to work with gender and with women because those of us who were leftist militants maintain a visionary belief in the struggle for better life conditions for the dispossessed classes" (1997, 71).

Bolivia's return to formal democracy, ensuing political-economic upheaval, and structural adjustments beginning in the mid-1980s initiated a period of uncertainty and redefinition for leftist opposition movements and set the scene for the emergence of multiple and sometimes competing currents among and inside NGOs. As social and political impulses rooted in class analysis began to refocus on questions of personal identity, changing alliances served to differentiate and oppose those who supported ethnic/peasant "others" (emerging as *indigenistas* or *andinistas*) from those who struggled for the rights of women (*feministas*), as well as to fuel factions within each current in which the *naming* of difference became a political battlefield.

As sympathies shifted, female middle-class leftists belonging to diverse parties began to define women as the group for whose rights they would fight. Participants in the Second Bolivian Feminist Congress in 1991 identified three tendencies: first, women's movements affiliated with political parties, which provided services, training, and various kinds of material and social support to women; second, professionals referred to as "institutionalized feminists," who worked through NGOs and internationally funded

projects to implement programs and advocate policies concerning reproductive health, domestic violence, economic opportunities, and related
issues; and third, scattered radical groups calling themselves "autonomous
feminists," who worked toward drastic social transformations. Sonia
Montaño, first subsecretary of Gender Affairs, acknowledged that the first
two groups engaged in a "necessary courtship" with the state as a strategy
to advance their goals (1993). In contrast, autonomous/anarchist feminists,
notably the La Paz-based group called *Mujeres Creando,* critiqued the middle-
class and heterosexual biases, financial dependence, and lack of ideological
autonomy of institutionalized efforts to advance women's interests.

Parallel to the rise of women's movements, men and women from various class backgrounds who had been active in prodemocracy movements
and leftist parties focused their efforts on support for ethnic identity and
rights for indigenous groups. Since the 1950s, the class label *campesino*
(peasant) had widely replaced terms connoting indigenous identity, while
mining and peasant unions increasingly nudged out traditional sociopolitical organizations. It seemed as if modern governance would render
Bolivia's indigenous politics obsolete. Then, "Almost unnoticed amid the
social upheaval of the mid-1960s, for the first time a generation of young
people from the rural hinterlands of both Peru and Bolivia began to find
their way to urban universities in large numbers" (Albro 1998, 102).
Among these new actors who would dramatically reintroduce ethnicity into
national political arenas was Victor Hugo Cárdenas, an Aymara who
became Bolivia's vice president in 1993.

The tendency of indigenous political and intellectual movements to
overlap is exemplified by the Andean Oral History Workshop (THOA),
who, since the early 1980s, has brought together indigenous intellectuals in
projects to document, study, and revitalize Andean social organization. As
Orta observes, "they find in the sociopolitical sensibilities of the *ayllu* an
opportunity for decolonizing Bolivian society and reimagining it as a pluri-
cultural space" (2001, 200). A recent article by THOA participants María
Eugenia Choque and Carlos Mamani (2001) demonstrates the symbiosis
between scholarship and movements through the description and analysis
of an experience in which communities in rural La Paz reorganized themselves according to symbolically powerful ethnic forms of participation and
leadership.

Middle-upper-class intellectuals have collaborated with highland and
lowland indigenous organizations in movements that vary considerably in
form and purpose, but generally share relatively traditional symbols of
indigenous identity. In startling contrast stand neopopulist phenomena
such as CONDEPA (Conciencia de Patria) and UCS (Unidad Cívica

Solidaridad), with their thoroughly distinct discourses grounded in urban indigenous culture, and exalting a kind of material success that is not synonymous with assimilation (see Albro 1998, 2000). It is notable that women, such as former La Paz mayor Mónica Palenque and presidential candidate Remedios Loza, have enjoyed a much higher profile in neopopulist politics than in indigenous movements or in traditional right- or left-wing political parties.

The dramatic emergence of this new political current brings into relief power/knowledge relationships embedded in more longstanding intellectual positions. Collaborations between social scientists and ethnic communities from the altiplano and the Amazon have given rise to important new approaches to territory and national identity that challenge key concepts and policies that have dominated the republic. Yet, for the most part, these efforts have not touched the everyday material existence and power relations of most urban men and women. In contrast, Bolivian feminist movements and researchers have questioned gender systems in elite institutions and family life as well as distant ethnic communities. Yet, partly because modern feminist projects are empowered by international discourses and funds, only self-consciously radical movements have interrogated key Western assumptions about gender roles and rights.

GLOBAL IDENTITY POLITICS AND LOCAL CONTRADICTIONS

The influence of global identity politics on the movements and conceptual currents sketched above has led to contradictory outcomes. While much anthropological research has emphasized the interwoven dynamics of gender and ethnicity in the Andes,[2] world-wide campaigns and declarations have tended to focus either on women and gender issues (e.g., World Conference on Women, Beijing 1995) or on indigenous issues (e.g., ILO Convention 169 on Indigenous and Tribal Peoples in Independent Countries).

Thanks partly to the efforts of international agencies during the late 1980s and 1990s, the notion of gender, heretofore an obscure academic concept, was brilliantly reincarnated throughout Latin America as part and parcel of modern development discourse. Many feminists working in NGOs and government agencies, together with feminist scholars such as Pozo and Tuijtelaars, initially embraced this modern gender discourse, partly motivated by the desire to achieve the respect (and funding) granted to rational modern thinkers, a status formerly denied them on the basis of gender. "Gender and Development" has so overshadowed gender theory's more scholarly origins that many Andeans sense that gender sprang, fully clothed, from the forehead of the United States Agency for International Development (USAID) or the Dutch Mission for Technical Assistance.

Individuals and groups focusing on ethnicity have not so quickly fallen in step with modernizing development. In contrast to the new field of gender studies, ethnic studies entered the development arena with a long tradition of ethnographic research in both Andean and Amazonian communities, and more recently, in urban contexts. This rich tradition has fueled ongoing theoretical and conceptual discussions, relatively independent of development politics and projects (although obviously embedded in the conceptual foundations of modern Western thought), and often deeply critical of them (as demonstrated by the Andeanist scholars presented above). So when people dedicated to studying ethnic issues began to work with the Bolivian government and other agencies to support and promote "ethnic" peoples caught in the throes of development, they brought their own methods and motives to the task. Unfortunately (or fortunately, depending on one's point of view), these well-established approaches, based on qualitative interpretations of complex cultural and historical processes, were not easily synchronized with reigning development paradigms and logical frameworks for project implementation. For example, the dispersed kin networks identified by Andeanist scholars above make unmanageable target populations for development projects, while spiritual connection with nature and culturally meaningful collaboration are difficult to isolate and measure as indicators of project success.

Although members of the respective subsecretariats of ethnic and of gender affairs shared political philosophies and analytic frameworks that should be able to speak to one another, they were supported by separate international agencies with funds often earmarked for one axis of work. When it came to defining objectives and implementing projects, the subsecretariats had trouble applying the idea of transversalización and ended up focusing primarily on actions to support marginalized others, defined in *either* gender *or* ethnic terms (Revollo Q. 1997). From its inception, the Subsecretariat of Gender Affairs was staffed mainly by women who researched women's issues and planned projects to support, educate, and empower women, in addition to coordinating with other agencies in efforts to incorporate women and gender sensitivity into their programs, specifically, local planning processes (SAG 1995, 1997). As far as I know, however, the subsecretariat has not carried out studies or projects concerning men's gender-specific demands or modes of participation in these planning processes, nor has it developed materials on other roles, practices, and problems of Bolivian men.

For its part, the Subsecretariat of Ethnic Affairs was staffed mainly by men and women who studied indigenous issues and designed policies and projects to support, empower, and defend indigenous groups. They also coordinated with other agencies to adapt national political processes to the organizational structures and dynamics of selected ethnic groups. Although individuals working in the Subsecretariat of Ethnic Affairs have in the past studied complex processes of interethnic mixing and shifting identities in contexts including urban centers, the Subsecretariat's work has prioritized isolated groups that are easily labeled as "other": Aymara ayllus of the altiplano, Amazonian tribes in the eastern lowlands, Afro-Bolivians in the Yungas.[3] Major Bolivian ethnic groups such as mestizos, urban Indians, cholos/as, or descendants of European immigrants fall outside of the subsecretariat's scope of work. While the subsecretariat has made valuable advances in self-governance of consolidated ethnic groups, emphasis on territorial autonomy fails to address the variegated landscapes through which people like Faustina Fernández move.

Turning the Tables: Marginal Groups Use Global Identity Labels in Struggles for Political Economic Justice

Many engaged Bolivians believed that the transversalización of gender-aware and culture-sensitive policies had potential to reshape important sites for the production of Bolivian identities. Yet the uncomfortable coupling of these initiatives with neoliberal economic programs has obscured their impact and ignited widespread mobilization against them. In Bolivia, like other countries that have undergone International Monetary Fund (IMF) imposed structural adjustments, multifarious struggles to shape society and to champion ethnic, gender, or class causes have been engulfed by a new type of development plan that prioritizes financial efficiency and economic growth. In fact, Bolivia's progressive reforms of 1993 to 1997 promoting grassroots political participation and cultural self-determination were anchored in a framework of structural adjustments that significantly eroded labor conditions, job security, and government-supported social programs.

Intellectuals and activists advocating gender and ethnic rights have developed ambivalent relationships with these politics, relationships that are variously interpreted as collaboration or prostitution, as becoming more sophisticated or selling out. Part of the seduction of neoliberalism is the promise of a kind of moral equality to be obtained by applying free market rules equally to all peoples. Against a history of political, economic, and legal systems explicitly based on class exploitation and on ethnic and gender inequality, such a universal principal can be tremendously attractive. Since the key agent and unit of analysis in liberal thought is the rational individual, neoliberal concerns with gender and ethnicity focus on turning individual women and Indians into effective economic actors.

Grassroots organizations, unions, and diverse social movements have been less ambivalent about the new state project, meeting structural adjustment policies—as well as accompanying cultural programs—with multiple forms of resistance. Bolivia's legendary miner federations and peasant unions, significantly weakened since the 1980s, have given way as new movements take the vanguard. A coalition of lowland indigenous groups joined in a historic March for Territory and Dignity in 1990, teachers' unions protested violently against education reforms in 1994, coca producers staged repeated marches and blockades throughout the 1990s, and a broad-based coalition arose to protest the privatization of Cochabamba's water supplies in 2000 and 2001. While holding tightly to ideas of class conflict that have long shaped resistance to government policies, these 1990s movements also engaged ethnic and gender debates in remarkable ways.

As they marched from their homes in the tropical Chapare region of Cochabamba to the distant highland capital of La Paz in January 1996, women coca growers (*cocaleras*) came to embody key ideas about indigenous and women's rights. For over a decade, the U.S. government had exerted pressure on its Bolivian counterpart by linking foreign aid funds to "certification," a label granted to those countries that collaborate sufficiently with the U.S. War on Drugs. To comply with standards, the Bolivian government implemented a series of harsh policies designed to eradicate coca plantations in the Chapare, including the controversial Law 1008 (*Régimen de la Coca y Sustancias Controladas 1988*), which establishes the coca-growing region as a military zone and overrides civil and human rights guaranteed in the Bolivian constitution.

The women's march to protest this situation became an intractable national conflict as the armed forces repeatedly intercepted the cocaleras, transporting them back to the Chapare three times before some women managed to follow rural trails all the way to La Paz (Lagos 1997). The march caught the attention of the national media and gained the support of opposition parties, the Catholic Church, and labor federations, even creating fissures within the government as some Congress members defended the constitutional right of cocaleras to engage in a peaceful march, while others feared that the movement would threaten conditions for fulfilling U.S. antinarcotics certification.

Government campaigns to discredit the march as illegal, politicized, or linked to international terrorism or narcotrafficking were contested by opposition rhetoric that skillfully echoed conventional paternalistic representations of women and indigenous peoples as helpless victims in need of government protection. A statement published by the Confederación Sindical de Colonizadores de Bolivia in the national newspaper, *La Presencia* (January 5, 1996), proclaimed that "The march of our *compañeras* constitutes a peaceful protest against the constant violation of human rights in the Tropics of Cochabamba... whose victims are principally women and children" (quoted in Lagos 1997). The testimonies of marchers themselves indexed global arguments for universal human rights, "We have walked for a month, under the rain, snow, and hail, along with pregnant compañeras, with hardly any food, and sleeping outdoors. We did all of this to make the government respect our women's rights and coca growers' rights."[4]

The most fascinating turn of the tongue came when the cocaleras redirected naming processes so as to impose gender and ethnic identities on people whose powerful positions normally made them immune to such labeling. Upon arriving in La Paz, the cocaleras demanded a summit meeting with the national powers, but rather than meeting with the president and vice president, they demanded to speak "woman to woman" with their wives. In a meeting broadcast on television, the cocaleras presented themselves as Quechua and Aymara women, foregrounding their sorority with the vice president's wife, who is Aymara. They emphasized their roles as mothers, wives, and sisters concerned with the repression that their family members suffered in the war against coca, and as victims of gender-specific violence, including sexual abuse, on the part of the military. A cocalera named Severina appealed to the president's sympathy, "Doesn't he have a wife, a mother, a daughter to know how difficult life is when one doesn't have anything to put in one's mouth?"[5]

The cocaleras' words and actions confound the polarization of gender and ethnicity that divides the scholarly community and the secretariat and challenge the state's effort to detour class conflict into discussions of individual identity and advancement. By emphasizing their gender, ethnicity, and nationality, the cocaleras strategically deployed those very dimensions of identity politics that are foregrounded in official discourse, going one step further by drawing their powerful interlocutors into these same identity categories (Lagos 1997). Yet, in contrast to the state, these indigenous women powerfully underscored the realities of their class position, and the profoundly unjust allocation of material, political, and military power that underlies the violent oppression of those distant marginal places that cocaleras occupy within the nation.

CONCLUSION

The research materials and analyses brought together here represent the merging of two currents that have shaped my personal voyage: that of late twentieth-century cultural anthropology and that of Latin American feminist theory. Anthropological fieldwork has long been about place, and as such has generated valuable insights on how gender and ethnicity work in diverse places of study. Engaged feminist research, transdisciplinary in nature, has mapped its debates on everything from individual women's experiences to universal human rights. Intersections between anthropology and feminism have sparked fertile tensions as anthropologists deconstruct purportedly universal gender categories and claims, while feminists interrogate the subjective positioning of scientific researchers and knowledge makers, including anthropologists, in campaigns demanding not only that we make the personal political, but also that we make scholarship personal. Together with other authors in this collection, I look to the restless borderlands of feminist anthropology for innovative strategies to explore gender's place, and to respond to what Comaroff and Comaroff identify as a central challenge of our era, "the problem of doing ethnography on an awkward scale, neither unambiguously 'local' nor obviously 'global'—but on a scale in between that, somehow, captures their mutual determinations. And their indeterminacies" (1999, 282).

This paper has brought together practices, discourses, and historical positions of people in different sites within Bolivian society in attempts to better understand the ways in which they come together and apart in ongoing struggles to define and enact identity. Unique relations of difference at play in Bolivia have clearly contributed to the emergence of the innovative visions and approaches manifest in government policies, in scholarly research, and in the actions and voices of indigenous men and women. Yet even as actors in the secretariat struggle to forge a more transversal place for gender and ethnicity in the political landscape, significant material inequalities and ideological barriers divide and constrain their efforts. Intersections between local movements and global trends have created similar paradoxes, or indeterminacies, in Bolivian intellectual communities. Meanwhile, institutions of power and knowledge (including media and development projects) continue to inscribe gender and ethnicity on the bodies of indigenous women, located far away from the powers that be. Finally, indigenous women move, chameleonlike, through manifold cultural spaces pervaded by gender and ethnic symbols, alternately eluding and appropriating for their own ends the labels and projects advanced by government agencies and scholars alike.

The intense relations of difference that gave rise to theoretically innovative concepts like transversalization seem to have also limited the potential of government agencies and scholars to apply these insights fully. At the same time, Bolivian dynamics have produced surprising situations in which less powerful groups claim for themselves the labels that have marked them as marginal others, appropriate international women's and indigenous rights discourses, and dare to publicly label members of the previously unmarked elite. The visions and actions of people like Faustina Fernández and the cocaleras continue to hold tremendous transformative potential.

NOTES

This paper took shape in the context of many animated conversations. I would like to give special thanks to Pamela Calla, coauthor of an earlier paper based on this research; members of my writing group at Miami University; and the three editors of this collection, all of whom provided valuable input. Some of the materials in this article were addressed in an earlier publication, "Gender and Ethnicity in Bolivian Politics: Transformation or Paternalism?" (Paulson and Calla, 2000, *Journal of Latin American Anthropology* 5(2): 112–149), which provides complementary analysis together with more extensive literature review. All translations are mine.

1. María Isabel Caero took part in an intellectual community constituted of sixteen women and men who were engaging gender ideas in diverse fields of work throughout Bolivia. A generous grant from the Dutch Embassy allowed me to bring the group together in workshops and to publish a collection of their work (Paulson and Crespo 1997).
2. See Arnold 1997 and 1998, De la Cadena 1997, Rivera C. 1996, Seligman 1998, and Weismantel 2001 among others.
3. Carmen Medeiros of the Subsecretariat of Ethnicity writes about the difficulty of addressing indigenous/ethnic issues in Andean regions where sociocultural units and corporate identities are not as easily identified as in the lowlands (Medeiros 1995).
4. Silvia Lazarte, Plaza San Francisco, La Paz, January 18, 1996, taped by Maria Lagos.
5. Severina, Ultima Hora, January 10, 1996, cited in Lagos 1997.

REFERENCES

Albro, Robert. 1998. "Introduction: A New Time and Place for Bolivian Popular Politics." *Ethnology* 37(2): 99–116.

———. 2000. "The Populist Chola: Cultural Mediation and Political Imagination in Quillacollo, Bolivia. *Journal of Latin American Anthropology* 5(2): 30–88.

Apffel-Marglin, Frédérique with PRATEC. 1998. *The Spirit of Regeneration: Andean Culture Confronting Western Notions of Development.* London: Zed Books.

Arnold, Denise, ed. 1997. *Más allá del silencio. Las fronteras de género en los Andes.* La Paz: Center for Indigenous American Studies and Exchange/Instituto de Lengua y Cultura Aymara.

———. 1998. *Gente de carne y hueso. Las tramas de parentesco en los Andes.* La Paz: Center for Indigenous American Studies and Exchange/Instituto de Lengua y Cultura Aymara.

Caero, María Isabel. 1997. "Espacios de género en el municipio de Cochabamba: Desencantos y esperanzas." In *Teorías y prácticas de género: Una conversación dialéctica,* edited by Susan Paulson and Mónica Crespo. Cochabamba: Embajada Real de los Países Bajos.

Choque, María Eugenia and Carlos Mamani. 2001. "Reconstitución del ayllu y derechos de los pueblos indígenas: El movimiento indio en los Andes de Bolivia." *Journal of Latin American Anthropology* 6(1): 202–224.

Comaroff, Jean and John Comaroff. 1999. "Occult Economies and the Violence of Abstraction: Notes from the South African Postcolony." *American Ethnologist* 26(2): 279–303.

Coronil, Fernando. 1992. "Can Postcoloniality be Decolonized? Imperial Banality and Postcolonial Power." *Public Culture* 5(1): 89–108.

De la Cadena, Marisol. 1997. "Matrimonio y etnicidad en comunidades andinas (Chitapampa, Cuzco)." In *Más allá del silencio: Las fronteras de género en los Andes,* edited by Denise Arnold. La Paz: Center for Indigenous American Studies and Exchange/Instituto de Lengua y Cultura Aymara.

Gill, Lesley. 1997. "Power Lines: The Political Context of Nongovernmental Organization (NGO) Activity in El Alto, Bolivia." *Journal of Latin American Anthropology* 2(2): 144–169.

Grillo Fernandez, Eduardo. 1994. "Género y desarrollo en los Andes." Lima: Proyecto Andino de Tecnologías Campesinas.

Gupta, Akhil and James Ferguson, eds. 1997. *Culture, Power, Place: Explorations in Critical Anthropology.* Durham, N.C.: Duke University Press.

Healy, Kevin and Susan Paulson. 2000. "Political Economies of Identity in Bolivia, 1952–1998." *Journal of Latin American Anthropology* 5(2): 2–29.

Lagos, María. 1997. "Bolivia La Nueva: Constructing New Citizens." Paper presented at the International Congress of the Latin American Studies Association, Guadalajara, Mexico.

Luykx, Aurolyn. 1998. *The Citizen Factory: Schooling and Cultural Production in Bolivia.* Albany: State University of New York Press.

Marcus, George. 1996. *Ethnography through Thick and Thin.* Princeton, N.J.: Princeton University Press.

Medeiros, Carmen. 1995. "Lineamientos para la elaboración del programa indígena de Tierras Altas." Manuscript.

Moffat, Linda, Yolande Geadah, and Rieky Stuart. 1991. *Two Halves Make a Whole.* Ottawa: Canadian Council for International Cooperation.

Montaño, Sonia. 1993. "El coqueteo necesario con el estado." In *La relación entre estado y ONGs*. La Paz: Cotesu/Ildis/Cooperación Holandesa.

Orta, Andy. 2001. "Remembering the Ayllu, Remaking the Nation: Indigenous Scholarship and Activism in the Andes." *Journal of Latin American Anthropology* 6(1): 198–201.

Paulson, Susan and Mónica Crespo, eds. 1997. *Teorías y prácticas de género. Una conversación dialéctica*. Cochabamba: Embajada Real de los Paises Bajos.

Paulson, Susan and Pamela Calla. 2000. "Gender and Ethnicity in Bolivian politics: Transformation or Paternalism?" *Journal of Latin American Anthropology* 5(2): 112–149.

Pozo, María Esther. 1994. "El riego desde la perspectiva de género." *Revista de Agricultura* 5(23): 307–314. Cochabamba: Universidad Mayor de San Simon.

Revollo Q., Marcela. 1997. "La experiencia de la transversalización del enfoque de género en la SNPP. Elementos para una evaluación general." La Paz-Bolivia: Secretaría Nacional de Participación Popular-Ministerio de Desarrollo Humano.

Rivera C., Silvia. 1996. "Trabajo de mujeres: Explotación capitalista y opresión colonial entre las migrantes aymaras de La Paz y El Alto, Bolivia." In *Ser mujer indígena, chola o birlocha en la Bolivia postcolonial de los años 90*, edited by Silvia Rivero. La Paz: Ministerio de Desarrollo Humano.

SAG (Subsecretaría Nacional de Asuntos de Género). 1995. *Cambiar es cosa de dos*. La Paz-Bolivia: Ministerio de Desarrollo Humano.

———. 1998. *Construyendo la equidad*. Ministerio de Desarrollo Humano. La Paz: Artes Gráficas Latina.

Seligman, Linda. 1998. "Estar entre las cholas como comerciantes." *Revista Andina* 16(2): 305–334.

Tuijtelaars de Quitón, Christiane, María Esther Pozo, Rosse Mary Antezana, and Roxana Saavedra. 1994. *Mujer y riego en Punata. Aspectos de género: Situación de uso, acceso y control sobre el agua para riego en Punata*. Cochabamba, Bolivia: Programa de Enseñanza e Investigación en Riego Andino y de los Valles (PEIRAV).

UDAG (Unidad Departamental de Asuntos de Género, Cochabamba). 1995–1996. "Diagnóstico complementario de género en la provincia de Tiraque." Cochabamba: Subsecretaría de Asuntos de Género.

Warren, Kay and Susan Bourque. 1991. "Women, Technology and International Development Ideologies: Analyzing Feminist Voices." In *Gender at the Crossroads of Knowledge*, edited by Micaela Di Leonardo. Berkeley: University of California Press.

Weismantel, Mary J. 2001. *Cholas and Pishtacos: Stories of Race and Sex in the Andes*. Chicago: University of Chicago Press.

7. The Racial–Moral Politics of Place ⌐⌐

Mestizas and Intellectuals in Turn-of-the-Century Peru

Marisol de la Cadena

INTRODUCTION

In the early twentieth century, Peruvian intellectuals tended to choose culture and its cognates (language, religion, and ethnicity) as the semantic fields to discuss the significance of race, and thus build its definition (de la Cadena 2000, 2001). By subordinating biological and phenotypic markers of race to the superior powers of morality and reason, elites from Peru—and other Latin American countries—could ignore skin color and biological hybridities. This conceptual strategy allowed them to safeguard their personal distinction from the threat of racial degeneration that North Atlantic thinkers—mainly from Europe and the United States—diagnosed as the reason for Latin America's political instability (Stepan 1980). The goal of my work has been to reveal the idiosyncracies in Peruvian intellectual definitions of race, and thus render visible the otherwise imperceptible threat smuggled in the apparently innocuous notion of culture, through which brutal exclusions and domination have been consensually legitimized. Accepting the challenge of the *compañeras* that edit this volume, my aim in this chapter is to continue *desalambrando* the concept of race. I interpret this assignment to mean writing against it (Gilroy 2000) by exposing its constitutive elements: the classificatory wires, *los alambres,* the taxonomic fences that contain individuals inside discriminatory hierarchies and the hegemonies that sustain them. In this chapter I work through dominant and subordinate definitions of the identity label "mestizo. " Locating my analysis in Cuzco (the capitol of the Inca Empire), my goal is to understand how these definitions, while emerging from competing racialized, classed, and gendered discourses, also coalesce in hegemonic forms of social discrimination. The task of desalambrar, while ultimately liberating, may reveal midway through that exclusionary fences are also built by those who are fenced in. Exposing this conundrum, while painful, is central to undoing discriminatory fences and freeing identities for all.

Culturalist representations of race were built upon images of intangible history—the soul of a people, its language, its spirit—which allegedly materialized through constructions, songs, crafts, and the lore of the indigenous (and in some cases Spanish) civilizations that had peopled the region. And "culture" dwelled in places. This is apparent in the following quote, written in the early twentieth century by one of Peru's most influential intellectuals and *Indigenista* writers, Luis Eduardo Valcárcel: "Each personality, each group, is born in a culture and can only live within it, like a fish in water. This universal relationship between human beings and the natural world is resolved through culture. We are the offspring, that is the heirs, of a being that has been shaped by the interaction of Nature and Culture" (1927, 109).

Sheltering nature and culture, "place" usually referred to as "environment," represented the soil and the atmosphere, where history acquired specificity and infused bodies with the inner characteristics that formed their racial soul. Place was to culturalist representations of race what phenotype was to its biologizing analogues: Not only did it shape cultures-races, it also marked them, and it did so in their innermost essence, their sexual behavior, their reproductive capacity. Once again, Valcárcel is illuminating:

> From a Freudian psychological viewpoint, these two natural regions [Coast and Sierra] would represent the two sexes. Femininity the Coast, masculinity the Sierra. The contrast was already visible in pre-Columbian times: people living by the sea cherished ease, comfort, and vicious pleasures, while those from the Andes were characterized by their roughness, their frugality, and their warrior-like spirit. ... [Today] Lima rejoices when visitors extol its femininity ... In contrast, Cuzco appreciates praises to its virility and pride. Lima is nostalgic for its Don Juan-like Viceroyalty; Cuzco has a longing for its sober kings, the Sons of the Sun. (1927, 118–119)

Costeños were different from *serranos,* and this difference was geographically inscribed in their sexualities. The moral backwardness of costeño gentlemen resulted from their feminine inclination to sexual pleasures.

In contrast, serrano men embodied a principled, stoic masculinity that was, in Valcárcel's view, the source of Peru's potential progress. At the core of this discussion was the issue of "mestizaje," defined at that time as "racial–cultural mixing." Should Peru become the fusion of Andean and Spanish cultures, like the Coast? Or should it, like the Sierra, emerge from its glorious Incan past? Neither question received an exclusive answer. Embraced by some and repudiated by others, mestizaje broached a question more than an answer. Mestizos, primarily defined as ex-Indians, former agriculturalists who had left the countryside for the city, were at the center of this debate throughout the twentieth century. Self-identified mestizos engaged in the debate as well. Among those, my focus is on the identity proposals presented by *Cuzqueño* market women, known locally as mestizas. In this chapter I analyze their ideas in political dialogue with dominant meanings that, while hardly seamless, drew primarily from the master narrative of *Indigenismo*.

A nationalist doctrine embraced by intellectuals from the Highlands and a few from Lima, Indigenismo anchored the Peruvian nation in its pre-Hispanic past, most specifically in the Inca legacy. Although many *Limeño* and foreign intellectuals would not have distinguished differences within Indigenismos, Cuzqueño elites could not ignore the variations between local "Indigenistas" and "Neo-Indianistas," as they broached opposing perceptions of proper elite male sexual behavior. Indigenismo—influential roughly from 1919 to 1930—preached a grandiose Incaic past, exalted racial/cultural purity, and despised mestizaje. Neo-Indianismo, which acquired prominence in the 1930s, exalted plebeianness as the paradigm for cultural authenticity and praised miscegenation. It followed that for Indigenistas, a morally proper masculine sexuality implied racial–cultural reproductive endogamy. Neo-Indianistas, by contrast, favored a notion of elite male sensuality attracted to both plebeian art and the eroticism of self-identified mestizas or *cholas*, as they referred to market women and *chicheras*, owners of working-class bars and experts at making corn beer (*chicha*). The two groups' appraisal of mestiza women also diverged: While Indigenistas portrayed them as the embodiment of the immoral sexual behavior that led to racial impurity and degeneration, Neo-Indianistas reinvented mestizas as the epitome of free women who, escaping the primitive, xenophobic sexuality of indigenous females, had creatively acquiesced to the solicitations of racial foreigners and thus become reproductive vessels of mestizaje. "The city" played distinct roles in each image: while for Indigenistas it signified the perversion of Indians (male and female), Neo-Indianistas thought of the city as a liberating venue for Indians, especially female Indians.

Self-identified mestizas broached an alternative opinion to both groups of elite intellectuals. I sought market women's viewpoints both in archives and through fieldwork in their stronghold, *el Mercado de San Pedro*. Redefining "place" has been central to this grassroots interpretation of mestizaje. First, mestizas counter the elite view of the marketplace as a locale where transgression and immorality reign supreme by portraying it as the setting of their struggle for subsistence, political recognition, and dignified identities. They thus deny portrayals of their sexuality as irrepressible and disprove the image proposed by Indigenista and Neo-Indianista elite intellectuals alike. Second, by presenting themselves as commuters, constantly moving between city and countryside, they belie the geographic determinism that confined "indigenous culture" to the countryside and doomed it to extinction in the city, where mestizaje terminally uprooted previous (rural and Indian) demeanors. These women, like other self-identified indigenous Cuzqueños/as, deny being "Indian," yet proudly enact indigenous culture in the city. "Indigenous mestizas," as some market women suggested I call them, have stripped the notion of mestizaje of dominant evolutionary connotations. For them, being mestizas does not entail shedding a previous (i.e., indigenous) skin; rather, it entails dressing that skin with civil rights and markers of a respectable identity.

"CADA UNO TIENE SU LUGAR" (EVERYONE HAS THEIR PROPER PLACE): DECENCY AND RACIAL PURITY

Indigenista regional and nationalist projects implied the legal defense of Indians (defined as a race of agriculturalists), the exaltation of Inca civilization, and the glorification of Cuzco as the "Cradle of Peruvianness," the sole origin of national authenticity. History, criminology, and archaeology were the disciplines that lent scientific support to Indigenista academic dialogues about Inca civilization and culture, which these intellectuals combined with tenets of *decencia* (decency) to construct a racial taxonomy that classified the regional population as *indios, mestizos,* and *"gente decente,"* decent people who were entitled to self-identification as *"blancos"* (see Corominas 1965).

"Decencia" was, and still is, an important component of distinction throughout Peru. In turn-of-the-century Cuzco, the ideology of decencia merged with notions of physiognomy in the production of intellectual racialist dialogues that could disregard the significance of phenotype in favor of a certain kind of refined morality and culture, capable of being inherited or achieved through education. Decencia allowed the local elite to secure their self-definition as "non-mestizo" in such a way that many among them could "have Indian features without being an Indian."[1] Broadly defined, decencia consisted in a reinterpretation of colonial classificatory orders of *calidad* and *pureza de sangre* in an era in which liberalism—and its egalitarian tenets—were supposed to prevail. Yet the ultimate meaning of decencia was "high morality," which could be ascribed by birth or obtained through education. As such, decencia was a powerful tool that could even modify original moral defects. Decencia informed people about local hierarchies, assigned them to their social places and, most importantly, kept them there.[2] Instilled by (class) education, decencia resulted in "propriety," the quality of respecting the social distribution of distinction.

The social antonym of decencia was *abuso*—abuse, the transgression of class rights and gendered place. Abuso reflected lack of education, the absence of moral restraints, and disregard for order—flaws deemed rampant among the populace where mestizos abounded. To elite eyes, the physiognomy of abusive commoners was strikingly disagreeable; they viewed it as revealing their dark inner beings.

Sexual behavior—considered inherently different between men and women and determined by the social and natural environment—was a cardinal point on the moral scale prescribed by decencia. As gente decente, the elites were sexually proper. Sheltered by the refinement of appropriately patriarchal urban life, *damas* (ladies) and *caballeros* (gentlemen) embodied untainted female virtue and the paramount male responsibility, respectively. In contrast, mestizas—exposed to the daily harshness and vulgarity of the marketplace or chicherías—were sexually aggressive females. Mestizo men, deformed in wretched rural villages where ignorance reigned supreme, were ill-fitted patriarchs, lazy and abusive merchants or landowners. Lastly, the indigenous population, though racially–culturally pure, was sexually frigid. They even lacked pubic hair, like the lack of vegetation in the cold altitudes where they lived. Brutish indios and sexually xenophobic indias completed the racial taxonomy of the region.

Indigenistas' sexualized racial–cultural taxonomy informed the anti-Hispanic nationalism that characterized their purist national project. This also found expression in the exaltation of an Inca mythic symbol of female virginal virtue. Cuzqueño thinkers identified four important women in the Inca pantheon: Mama Huaco, Chañan Cori-Coca, Cori Ocllo, and Isabel Chimpu Ocllo. The first two were warriors: Mama Huaco had saved her lineage in mythic times and Chañan Cori Coca had fought the Inca's enemies, the Chancas, in historic Inca times.[3] The last two were Inca noblewomen who witnessed the Conquest: Isabel Chimpu Ocllo, who represented the origins of mestizaje was the mother of the famous chronicler Inca Garcilaso de la Vega and the prototype of the indigenous princess who surrendered to the conqueror and was subsequently abandoned by her Spaniard love.[4] Cori Ocllo, in contrast, was the Inca noblewoman who preferred to die rather than surrender to the conqueror, and represented the preservation of indigenous racial purity. Cori-Ocllo's image was perfectly suited to the 1920s Indigenistas' aspiration to racial purity. Importantly, it was compatible with the elites' self-prescribed racial/cultural endogamic sexuality. Luis E. Valcárcel, who had her statue preside over his office at the *Instituto Histórico del Cuzco*, wrote Cori Ocllo's story, noting how "Don Gonzalo wanted her for himself, and she was faithful to her race. How could she offer her body to the impure assassin of her gods and of her kings? She would die first; so she lay tranquilly, without further vexation; to her cold flesh the white beast would not dare come close. … Kori Ojllo [*sic*] in order to frighten away from her the Spanish gallant had covered her perfect torso with something repugnant capable of driving away Don Juan himself. But still more virulent was the hatred that her eyes distilled." He emphatically ended his depiction sentencing: "The impure Indian woman finds refuge in the city. Flesh of the whorehouse, one day she will die in the hospital."[5]

Although mythic, the image of Cori Ocllo also fit with different kinds of eugenicist projects—and particularly with anti-immigration policies that opposed racial mixture based on the argument that sharp sexual differences existed between "more developed races" and Indians.[6] In the midst of their primitiveness and sexual frigidity, however, Indian women were crucial in maintaining racial–cultural purity: Like Cori Ocllo, the Indian woman would allegedly die before surrendering to a racially foreign man.

Their xenophobic sexuality also neutralized female Indians as objects of gentlemen's sexual desire. Refined as they were, they did not find pleasure in arousing female Indians' frigidity. Instead they wrote poems that praised the purity, nobility, suffering, and virginal chastity of Indian women.[7] Yet, to spark this kind of intellectual admiration, Indian women had to remain rural. The rural community or *ayllu* (allegedly organic to Indians' agricultural nature and history) purportedly sheltered Indian women from sexual assaults. But this protection did not reach the city. There, Indian women suffered the assault of immoral mestizo men, former Indians psychologically perturbed by migration and the consequent ill-adjustment to city life. The urban Indian woman—the mestiza—having abandoned her proper racial place, was condemned to poverty, violence, and illness. A fish out of water, in the city she would gradually degenerate morally and sexually. The "impure Indian woman"—as Valcárcel called those who did not follow the example of Cori Ocllo—ended as a prostitute or jailed as a thieve. Cori Ocllo's image was the emblem of Indigenistas' repudiation of mestizaje and, significantly, of the mestiza market women.

These successful and bold working class urbanites became the targets of the modernizing project of the 1920s and its moralizing tenets. During these years, Cuzqueño rulers launched a series of sanitary campaigns (see Guy 1991; Findlay 1999). The marketplace—mestizas' territory and the site of undesirable, yet inevitable, encounters between them and gente decente—became the obsession of the cleansers. A dangerous setting from which moral and material filth could spill into the city, the marketplace required tight control and domestication. The physical conditions of the locale were certainly deplorable: The lack of running water turned the bathrooms and the entire marketplace into a source of contagious diseases. However, in addition to cleaning the locale itself, the cleansing project targeted the allegedly dirty mestizas. Municipal authorities deemed that the meat, fruit, and vegetable stalls owned by these women were a health threat: They attracted "large quantities of flies that conveyed millions of germs" (Arguedas 1926, 1928). Even mestizas themselves—their filthy appearances—were a constant source of worry to city rulers. They proposed banning the "*vestido de castilla,*" a colorful woolen skirt with which mestizas marked their identity as self-respecting and successful indigenous urbanites. To the elite male authorities, however, the skirt was a "favorite nesting material for bugs and filth of all sorts, a permanent carrier of bacteria." Instead of their usual dress, mestizas were to wear white, body-length aprons. This color would facilitate the supervision of their cleanliness. It was additionally proposed that the market women be obliged to wear short hair, and not the typical long braids—another important marker of self-identified mestizas—which hygienists argued frequently came in contact with the produce they sold, contaminating it (Arguedas 1926, 1928). Applied energetically, the municipal laws regulated—and therefore rendered as transgressions—existing confrontations between the authorities and the market women. Newspapers repeatedly bombarded their predominantly elite readers with information about the vulgarity and filthiness of *placeras,* the "women of the plaza," as they were disrespectfully called:

> Yesterday a woman called Rosa Pumayalli, a meat seller, set off a serious disturbance in the marketplace because the guard ... informed her that she had to wear her apron and clean her belongings in conformity with the orders from the mayor's office. That was enough for Pumayalli, whose Quechua name means "vanquisher of tigers." She tried to attack the guard and chased him brandishing a knife and showering him with insults from her well-supplied repertoire. ... This sufficed to encourage the rest of the butchers to rebel against the Municipal decree concerning tidiness and cleanliness. ... It would not be strange if one day we had an armed strike in the Mercado Oriental where there is an abundance of dirty and indecent Pumas who do

not want to remove their secular crust of filth. (*El Sol,* September 9, 1926, 3)

Although Cori Ocllo might have not been in the mind of the journalist who wrote those lines, her image was the antithesis of these mestizas. Robust to the point of engaging in physical confrontation with two guards, bullying everyone with their obscene vocabulary, these women were the embodiment of material and moral filth, and targets of the overt disdain of "gente decente." Certainly, their dislike for mestizas (and mestizaje) represented a conservative class rhetoric directed against an incipient commercial bourgeoisie that was emerging from among the popular classes and that threatened to control urban life through the crucial control of the marketplace, the "stomach of the city," as local people referred to it.

NEO-INDIAN CHOLOS, VIRILITY AND MESTIZAJE

Indigenistas lived and exercised their authority in Cuzco in the 1920s, the period that broadly coincides with the modernizing government of Augusto B. Leguía. When the president was ousted in 1930 by a military coup, the most prominent Indigenistas, ardent opponents of the ex-president and collaborators in his overthrow, migrated to Lima and joined the new government. Although the Indigenistas continued to be influential in Cuzco, their physical absence coincided with the emergence of populist political ideologies that favored the emergence onto the cultural scene of an iconoclast and populist (yet elite) intellectual group. Opposed to Indigenistas since their inception, the new group identified their ideas as Neo-Indianismo (Neo-Indianism). Prominent in Cuzco between World War I and World War II, they joined the international antirace movement[8] and repudiated their predecessors' project of racial purity. Like Mexican Indigenismo, their nationalism championed a mestizaje whose final stage they defined as a spiritual fusion of cultures, to be attained through the promotion of plebeian art, which they called folklore.

From their positions in official city institutions they promoted "folkloric contests" and fashioned their bohemian personas, publicizing their ability to produce "plebeian" art themselves. Poets, guitar players, and dramaturges, their highest achievement was the 1944 creation of the *Día del Cuzco,* a day to celebrate their city. They also mounted the first official production of *Inti Raymi,* a reinvented Inca festival that continues to attract Peruvian and foreign audiences to this day. With folklore as their nationalist technology, the Neo-Indianistas sought to extract the essences necessary to surmount racial–cultural differences, and to forge a common mestizo soul. Born of the virile stamina allegedly deposited among the working-classes, and most specifically in the *cholo serrano*, the mestizo soul became the symbol of the future regional identity. Described by a journalist as "más bajo que alto, más obscuro que claro, más cholo que indio, más indio que mestizo, más mestizo que blanco,"[9] (shorter rather than taller, darker rather than lighter, more cholo than Indian, more Indian than mestizo, more mestizo than white) images of cholos—a label referring to working class men—explicitly slipped away from any phenotypical definition, yet leaned toward the most plebeian side of the urban social scale.

Crucially, as Neo-Indianistas emphatically blurred the racial component of the future cholo, the one that only elite bohemian men—like themselves—embodied, they emphasized his masculinity. In their project virility was the essence of nationality: it united "authentic" Peruvian men, defined as those who dared to feel attracted to the unrefined culture of the populace. Virile authenticity expedited the Neo-Indianistas' hierarchical description of regional identities. Indians were future cholos, but were not quite cholos yet. Although their ability to survive in the most hostile environments evinced their masculine strength, Indians' artistic coarseness indicated a still primitive sensuality (including their sexuality) that needed to be disciplined. Once this was achieved, they would be neo-Indios. Allegedly democratic and anti-racist gentlemen, neo-Indianistas saw themselves as paradigmatic mestizos. They self-identified as cholos, and boasted of their desire to transgress the social order and share a place with "the people." The chicherías that Indigenistas visited surreptitiously became the unrestricted meeting places of these bohemian gentlemen.

CHOLAS AND NEO-INDIANISMO

A crucial feature of Neo-Indisanista's identity was their public declaration of the attraction *that cholas (and particularly the chicheras) exerted on them. This declaration seasoned their* cultural–political project, their revaluation of mestizaje, and their idiosyncratic perception of elite men's sexuality. Proud of their taste for all things plebeian, they exhibited their rough sensuality by expressing their inclination for folk art and proclaiming themselves to be unparalleled sexual partners of cholas. Needless to say, in Neo-Indianistas discourse the virginal image of Cori Ocllo lost ground. Instead, Isabel Chimpu Ocllo, the mother of Garcilaso de la Vega represented the pinnacle of Neo-Indianismo. Known as "the first Peruvian mestizo," and the same woman whom earlier Indigenista mythology had accused of sexual capitulation, according to neo-Indianistas Isabel Chimpu Ocllo was:"... the docile woman who served as amiable solace when the conquistador was an orphan of love ... America overcoming its tragedy. Isabel is the American cave that indigenous mythologies call '*pakarinas*' [places of origin] [*sic*], from where the makers of the new culture arise again. From her entrails, as from these marvelous grottos from which the Incaic demiurge emerged, the undomesticated mestizo is born, [this] will give a renovated vigor to the Continent" (García 1937 [1930], 142).

Under the neo-Indianista influence, mestizaje ceased to signify indigenous women's racial/sexual treason, representing instead a fertile inclination toward progress. Having surmounted the sexual xenophobia inherent to their race, these women were no longer Indians; they were mestizas, more commonly known as cholas. "(T)he chola is the rejuvenated organic force that advances impudent and fearlessly toward the city and toward the present, nurturing with her abundant and maternal breasts the energy of the race, like a mother or a wet nurse, smelling of chicha, and with a huayno in her throat" (García 1937 [1930], 182). Exuding sensuality and ubiquity, the chola had the power to transform places. She aroused the dormant rural areas as she was the "flame that ignites with passion this countryside" (García 1937, 212). In the city, she displayed her progressive commercial ability (and coarseness) as a market vendor, while reserving her irrepressible—and irresistible—eroticism for chicherías, the cradle of the *Nuevo Indio* and nuptial bed of his parents: the chola and the bohemian populist gentleman.

Neo-Indianista preaching about the chola was echoed nationally and internationally. In Cuzco, the cholas' imputed sensuality inspired poems that sharply contrasted with those the Indian woman's alleged virtue provoked. In 1944, to celebrate the first *Día del Cuzco*, the famous local poet Luis Nieto wrote the lyrics to *Himno del Cuzco*. The same author also dedicated the "Romance de la Barbaracha" to the cholas of Cuzco, an excerpt of which reads,

> Chola I love you as a chola
> fresher than an apple
> with your round skirts
> and your blouse of percale.
> Your smile shines clean
> as the crystal of the hoarfrost.
> Your eyes shine mischievously
> the sun plays on your lips,
> the moon envies your face.

Your two braids to me seem
two vipers along your back.

...

I wish I were a hawk
to seize you with my claw
and to behead in your bosom
your two slave doves.
Nobody loves you as I do
bold and pretentious chola
with your fourteen skirts
and your pure wool shawl

...

little chola, very pretty chola
lift your skirt a little.
They have told me that your thighs
are rosier than the dawn.
Show me how you like
to furiously move your buttocks
They make me want to sack them
with my pirate hands! ... (1942, 25)

Neo-Indianista populist mestizaje was a stratified masculine project in which cholas chicheras were the sexual ideal of Neo-Indianista men, the cholas' conquerors. Although Neo-Indianismo encouraged the transgression of norms prescribing elite men's public sexual endogamy, their masculinity yardstick maintained other norms of decencia, including men's patriarchal duty to shelter women from the hardships of wage earning. Certainly Neo-Indianistas praised the joy, strength, and artistic inclinations of plebeian cholos. However, when it came to evaluating his deportment as the chola's partner, the cholo emerged as "a lazy tramp [who] exploits her and forces her to work with blows" (García 1937 [1930], 118). Notwithstanding the cholos' carnal sensuality—which accounted for their unquestioned virility—and artistic inclinations, their lack of manly responsibility subordinated plebeian cholos to Neo-Indianista cultural leaders. Neither effeminate aristocrats like their predecessors, nor irresponsible like working class cholos, Neo-Indianistas had the virile potential to subjugate cholas and to publicly declare their sexual deeds, qualities that placed them at the cusp of regional masculinity.

Despite the relative popularity among elites of the cultural/historical aspects that Neo-Indianismo promoted, the grotesque figure of the "*mujeres del pueblo*," the *chola vulgar* that early Indigenistas despised, remained prevalent in Cuzqueño upper classes' minds. Yet self-identified mestizas had an alternative code they called *respeto*. A far cry from decencia, the moral foundation of respeto was not built upon sexual propriety but rather upon a work ethic that defied the sheltered comfort of a lady's life. It is from this code of respeto that mestizas have forged an alternative self-image—one of a working class mestizaje that disrupts distinctions between city and countryside and appropriates the urban marketplace as the site linking the two.

THE MARKETPLACE: POLITICAL STRONGHOLD OF MESTIZAS DE RESPETO

In 1944, the market women joined trends set by the new populist political parties and created the *Sindicato de Mercados Unidos* (United Markets Union) (Aranda and Escalante 1978, 49). The Sindicato was to become one of the most active and regionally important among the many unions that mushroomed in both the city and countryside by the mid-twentieth century. By the 1960s, the unions were successfully leading the massive seizures of hacienda lands that would culminate in 1968 with the radical Agrarian Reform (see Rénique 1991). Hacienda invasions were not isolated countryside events. Rural political leaders demonstrated in the *Plaza de Armas*, where urban leaders joined them during region-wide general strikes. There was no successful strike without the participation of mestizas, who contributed to paralyzing the city by closing the marketplace. Through such actions, mestizas changed the image of the marketplace from a place where filth reigned and infections diseases were propagated to a symbol of the popular classes' struggle, the political stronghold of the people. "Scandal"—the subversion of the order imposed by decency—started in the marketplace from where mestizas and their families marched en masse to the Plaza de Armas shouting slogans and delivering speeches against urban or regional authorities. "Scandal" deeply perturbed ladies and gentlemen: The sight of women fighting for their rights and, even worse, protecting their men offended the moral code of proper patriarchy, belied the authorities ability to control the populace, and was, thus, a very effective political tool, one that the elites feared and could not tolerate.

A wonderful example of the effectiveness of market women's political action is a particular General Strike that lives in Cuzqueño working-class memory as the most glorious moment of the Federation of Cuzco Workers. Called in response to a rise in the price of gasoline, the strike was particularly important to market women because fuel was a key part of the cost of transporting harvests and cattle from the countryside to their selling stall. Witnesses recall the important role mestiza vendors played in the event: "The police entered the marketplace violently, and the vendors reacted, repelling the aggression ... some moments later the women captured

General Daniel Vargas Dávila … six female market vendors participated in his kidnapping; they grabbed him by the neck, pulled off his military jacket and his hat, and urinated in it" (Aranda and Escalante 1978, 61). Recalling the event, Agustín Mamani, a leftist politician and the Secretary General of the United Markets Union during the riot, mentioned that the strikers prevailed when a group of market women captured the local commander in chief of the army, whom they handed to Agustín. "We took the General to the locale of the Drivers Union where we forced him to negotiate an end to the strike. He was so surprised at the mestizas' courage that he would not offer any resistance."[10] Agustín did not mention that the women had urinated in the general's hat. Whether or not this happened does not matter that much, for the sight of a group of women kidnapping a general of the Peruvian Army may have been as shocking as the women urinating in his hat. Both were conflated in the annals of popular resistance, where "scandal" was and continues to be a powerful reversal of the rules of decencia and the imposition of mestizas' respeto.

From the mestizas' point of view, their alleged "moral defects," their insolence or *atrevimiento* (which means insolence but also boldness) were (and still are) expressions of the courage with which they defended their rights as members of the working class and as mujeres del pueblo. In this sense, market women have forged their counterimage as *mujeres que se hacen respetar* (women that make people respect them). Lucrecia Carmandona— a mestiza marketwoman from whom I learned a great deal, described this to me:

> I have stood up to anyone who failed to respect me. I have insulted them and chased them out with knife in hand. I learned to talk to defend myself and I have done it. … I do not belong to the high society of Cuzco, those who despise us. They call us "those cholas," they insult us; they think we are thieves and whores. I am only a worker and I have helped my husband all my life. if I have defended myself and insulted them, it is for my children. Because of that all my sons have been educated and now all of them, every one is a professional … but my struggle continues because now I still have grandchildren who have to grow up.

This image was already at work in the 1920s, as the earlier quote about Rosa Pumayalli illustrates. Women like Lucrecia and Rosa epitomize the plebeian mestizaje project and its relentless striving for respeto. Rejecting categorically the label of cholas, which taints their identity with allusions to an uncontrollable female sexuality, they have developed their own interpretation of moral behavior built upon the values of "honorable work" and "endless boldness" to protect their families. They describe their behavior as seeking respeto and promoting their children's formal education. Because fearlessness, the utmost female defect according to gente decente, is central to this image, these women continue to provoke rejection among Cuzco's elite—despite some mestizas' economic success. Thus the elite have given one of the most successful mestiza entrepreneurs, a tomato wholesaler, the scornful nickname of "the queen of tomatoes." Indignantly, they note how this mestiza "even bought a colonial residence that belonged to a Cuzco aristocrat" who had fallen on hard times. (I was unable to confirm the veracity of this story.) Mestizas' rampant and successful defiance of elite decency dismisses its gendered racial taxonomy and proposes respeto as an alternative moral code. This alternative code, however, is not free of its own hierarchies and forms of exclusion.

INDIGENOUS MESTIZAS AND THE
HIERARCHIES OF RESPETO

Uriel García—the most visible champion of Neo-Indianismo and an admirer of mestizas' eroticism—identified clothing as the essential marker of the change from Indian woman to mestiza identity. Comparing the process of changing clothes to a natural metamorphosis, he wrote that the mestizas' skirts were the "chrysalis of the Indian woman who becomes mestiza" (1949, 185). There is an obvious discrepancy between the elites' interpretation of this change in attire and the interpretation given to it by Lucrecia and other market women with whom I conversed. From the perspective of both Indigenista and Neo-Indianista intellectuals, changing clothes from "Indian woman" to "mestiza'" connoted a terminal movement from one racial–cultural group to another, a process intimately marked by the acquisition of a new sexuality. For the mestizas, being an Indian or a mestiza is not about racial *or* sexual differences. It implies "changing clothes," an action that, as in the elite version, expresses a social transformation. But this transformation does not require shedding previous cultural baggage. Rather, it may entail acquiring the means to better express and live regional indigenous culture.

Thus, even as the ethnocategories "Indian" and "mestiza" connote educational differences, these do not involve the evolutionary process from primitive to civilized suggested by the dominant notion of mestizaje. Nor do they connote the existence of cultural/geographic boundaries separating (rural) Indians from (urban) mestizos. Instead, from an indigenous viewpoint, the difference reflects a social relation inscribed in power. Mestizas command authority in relation to Indians because, *unlike the latter,* they earn a respectable income, they are not servants, and they have acquired the skills to live in the city. Mestizas are high-ranking indigenous women *in relation to* those they deem Indians—but they know that they are also Indians vis-à-vis gente decente, and therefore their taxonomies are infinitely relational.

The following excerpts from Lucrecia Carmandona's life story illustrate this relational transition—as well as her constant struggle to avoid misrepresentation as Indian, if only because of the civil wretchedness this condition implies. Lucrecia told me that when she was an adolescent, she had been an Indian servant. As such, one of her tasks was to sell the produce from her employer's small plot in the marketplace of the nearby town. As she learned to sell, she gradually learned how to be a mestiza. She bought a white hat ("the typical mestiza hat," she said) to protect her head and face from the sun and acquired an apron to keep her skirts clean "like all of us market women have to do." "I wore none of this when I was in the house; they did not want a mestiza as a servant, they wanted an Indian. My employers would not let me wear this, I think." When she was fifteen and "already knew how to sell and could be a mestiza," Lucrecia escaped from the house where she worked as a servant. She went to Cuzco, bought a basket and a few bananas, and began selling. "Walking up and down the streets with my basket, my apron, and my hat, I was still learning how to be a mestiza. I was not one yet, and people did not consider me totally one. I was not a servant anymore, but I was still working in the streets." Lucrecia concluded, "Now people consider me a mestiza, a very good mestiza. I have a permanent stall in the marketplace, I have good clothes, but I had to learn a lot before this."

Throughout her story Lucrecia emphasized the process of *learning* how to wear clothes, and constantly measured herself against the standards of a "good mestiza," the cusp of female indigenous identities. When she reached that cusp, she became a full participant in Andean practices: "When I was a servant, I could never go to *Coyllur Rit'i*, it was too expensive for me." She then noted, "I have gone several times lately." *Coyllur Rit'i* is an indigenous pilgrimage and an occasion when so-called Indians and self-identified mestizos convene to pay ritual homage to one of the sacred landmarks of the region. As an Indian servant Lucrecia yearned to go to *Coyllur Rit'i* yet could not, for she had neither the money nor the occupational freedom to do so. As a mestiza vendor she acquired the liberty to straddle city and countryside and could afford the expenses implied in the pilgrimage: She was not an Indian anymore, she was an indigenous mestiza.

By scrutinizing differences in the quality of the fabric, the type of shirt or blouse, the way of wearing the shawl and the material of the hat, urban indigenous Cuzqueños have ranked indigenous women along an ethnic continuum, with the "Indian women" and the "*buena mestiza*" occupying inferior and superior ends, respectively. In addition to their attire, the produce they sell contributes to the mutual classification of female vendors. Indigenous female buyers and sellers are classified as Indians (or *mujercitas,* "little women") "*mestizas simples*" (simple mestizas) "*buenas mestizas,*" (good mestizas) and "*casi damas*" (almost ladies). Nonindigenous buyers (almost never sellers) comprise "*casi damas*" and "*damas.*" This ranking allows for the situational interpretation of interactions. Teófila, for example, is a fruit seller who identifies her mother as a *mestiza simple,* and herself as a *casi dama.* As she explained,

> *Mujercitas* never eat chicken; on special occasions they buy *menudo* [beef parts] for their parties. In February when their own harvest runs out they buy handfuls of potatoes. They do not buy anything else in the market. *Mestizas* buy second or third category beef meat, maize, wheat, *chuño* [freeze-dried potatoes], potatoes, onions. *Damas* buy chicken, first-rate meat, potatoes and fruit of the valley. "*Gringos*" [an ethnic category introduced after the Agrarian Reform that includes all foreign workers of the development institutions] buy lots of fruit, vegetables and chicken.

A peculiarity of this classification not apparent from the above quote, is that indigenous Cuzqueños do not have individually fixed identities, but identities that are relative to others and contingent upon the status of the persons in the interaction. This is not a new phenomenon. For example, in the early twentieth century, Polonia Arnao, a woman from the nearby town of Yucay, insulted her brother after a domestic quarrel over their father's property by calling him "*Indio hereje ladrón de la casa de tu padre,*" (heretic Indian, a thief in your own father's house) while she self-identifed as mestiza. The same Polonia, in the broader scenario of her trial in the urban courts of Cuzco, called herself an Indian, probably to claim the "favors" granted by the Law to "semi-civilized" Peruvians, but also because the judges would willingly classify Polonia as such.[11]

Situations like this negate the idea of racial–cultural transformation from Indian woman to mestiza that Cuzqueño intellectuals used to define the "fixed" identities that their evolutionary, boundary-making taxonomies needed. When one takes into account the inherently relational quality of Cuzqueño indigenous identities, the boundaries disappear. Although the lexical classification remains intact, "Indian" and "mestizo" identities shed the rigidity they have in the elite taxonomy and acquire instead a processuality that makes *both of them* hybrid in a Bakhtinian sense: They are simultaneously different and similar, and they inevitably evoke each other.

As in Polonia's case, it is not uncommon for indigenous Cuzqueños to self-identify as both Indians and mestizos, depending on the relationships involved in a particular social situation. Likewise, symbolic actions can alter identities in one interaction. For example, when the municipal police wanted to offend market women, they pulled their hats or shawls off, as both articles identified them as mestizas. By removing these markers the police implied that they had the power to recast them as Indian women.[12] Tensions ensuing from such ethnic/social hierarchies pervade daily life in the *Mercado Central,* where demanding "respect" can also mean demanding "better treatment" and a better place on the marketplace social scale. Conversely, "to show disrespect toward someone" means to diminish their social place, often in a subtle ethnic idiom. Interactions in the marketplace are impregnated with considerations about the social identities of the parties involved in the relationship, and insults generally mean an assault on ethnically identified social status. However, harassment is not always considered "abusive," since the treatment a person merits depends on how his or her behavior is evaluated, according to his or her social condition. On one occasion I witnessed a vegetable seller scolding a porter—the lowest in the scale of male Indianness in the marketplace. She told him in Quechua, "dirty, lazy Indian, you should have carried the basket like I had told you to, it was your obligation to do so." When I asked the other witnesses— mostly market women—about their colleague's behavior, they told me that since the porter had refused to work for the vendor and carry her basket to a hotel, it was the woman's right to scold him. "That señora is nice; when they do what she commands, she even gives the porters lunch," some women agreed. They thought the porter was receiving the treatment his Indian "insolence" deserved. Ironically, the behavior of market mestizas reflects patterns of social distance that permeate the social construction of ethnic identities in Cuzco.

However, nothing remains fixed in the Mercado Central where daily confrontations encroach on established hierarchies, and where trespassing them is a norm. Ladies constantly complain that mestizas are more uppity than ever; while mestiza vendors say that porters "get more deceitful every-day. They don't work the way they used to anymore."

CONCLUSIONS

Indigenismo described mestizaje as the precarious misplacement of Indians (racially agriculturalists, and obviously rural) in the city. Neo-Indianismo reinterpreted mestizaje as a powerful symbol of cultural integration, populist coalition, and national identity. Both creeds stressed the hybridity of the "new" person, and opposed it (positively or negatively) to a racially or culturally pure Other. Meanwhile, indigenous working classes (who are neither exclusively urban nor rural, but Andean-style commuters) partially dismiss this dominant notion and advance a different one.

From an Andean viewpoint, mestizas are successful indigenous women, equivalent to the Aymara women of La Paz, who fuse city and countryside, thus canceling the implicit dichotomy that defines rural and urban as discrete spheres (cf. Gill 1994). By calling themselves mestizas/os, urban indigenous Cuzqueños rebuke colonial stigmas and present an image of success that removes indigenous practice from the social condition of Indianness and its implied historical wretchedness. When this aspect, along with the inherently relational aspect of Indian and mestizo identities are taken into account, the "changing of clothes from Indian to mestiza" reflects changing social conditions but does not suggest that "cultural passing" results from migration to the city or from acquiring an urban job. The differences between mestizas' and Indian women's clothing, whether striking or subtle, convey power differences usually accompanied by a better economic condition. Viewed from this perspective, the boundaries between Indians and (indigenous) mestizas result from social differentiation and are not necessarily "ethnic." Hence, becoming mestiza/o implies distancing oneself from the Indian social condition, and is therefore a decolonizing process, a far cry from "disappearing" into a homogeneous, nationally shared, mestizo culture as purported from a dominant viewpoint. Rather than "evolving" and moving away from Andean culture, mestizas are indigenous woman that demand respect.

In Cuzco, de-Indianization is the process of empowering indigenous identities through economic and educational achievements and displaying this identity in regional events of popular culture that take place ubiquitously in urban and rural stages.[13] This ubiquity disrupts the evolutionary relations between place and identity. It abides by the agreement that indigenous culture originates in the countryside and accepts the relocation of successful Andeans to the city, thus contradicting the dominant belief that fixes indigenous culture in the countryside, and thus enables decolonizing strategies. Yet de-Indianization also harbors jarring situations. Respeto, the underpinning of assertiveness, identifies superiors and inferiors in a relationship. Although superiority and inferiority are defined in face-to-face relations, once the parties involved reach an agreement about who is who in the interaction, respeto allows for discriminatory practices within indigenous working classes. Thus, mestizas themselves live in the grip of Cuzqueño discriminatory practices. "Respect" means observing the social hierarchies that market women and men recognize, and although these abide by alternative working-class moral codes, they also coincide with some of the hierarchies imposed by decencia. Rather than the prerogative of the elites, domination is the discontinuous result of constantly negotiated consensus among diverse—and internally differentiated—social sectors.

NOTES

My biggest *agradecimientos* go to Janise, Lessie Jo, and Charo, the editors of this volume. Their care, thoughtfulness, and passion have inspired me beyond the rewriting of this article. *Gracias* also to Susan Paulson; I hope to have lived up to her wonderfully insightful comments.

1. Luis E. Valcárcel (1981, 68) used this phrase to describe a famous Cuzqueño priest.
2. Mariscal 1922, 12. Archivo Departamental del Cuzco (ADC), Universidad Nacional de San Antonio Abad del Cusco (UNSAAC), Libro N. 16.
3. Cornejo Bouroncle 1949, 7–9.
4. Cornejo Bouroncle 1949, 17–29.
5. Luis E. Valcárcel 1978 [1928], 78.
6. See Escalante 1910. On the intellectual predicament of European colonialism, hybridity, and sexuality, see Stoler 1994; Young 1995.
7. See for example Masis 1927; Lira 1960.
8. On the postwar rejection of race as a valid concept in the Northern Hemisphere see Stepan 1982 and Barkan 1992.
9. *El Sol,* November 6, 1946, 3.
10. Agustín Mamani and Lucrecia Carmandona de Mamani, personal communication. See also José Sotomayor Pérez (1984).
11. ADC, Fondo Oscar Zambrano, Legajo 209, 1934.
12. AHMC, Legajo 167, 1957, November 2, 1957.
13. This redefines previous scholarly definitions of de-Indianization that did imply assimilation and acculturation. An important example of this position is provided by Guillermo Bonfil Batalla 1996. See also Friedlander 1975.

REFERENCES

Aranda, Arturo and Carmen Escalante. 1978. *Lucha de clases en el movimiento sindical Cusqueño 1927–1965.* Lima: Herrera Editores.
Arguedas, Luis A. In *El Sol,* February 3, 1926, 3 and *Revista Mundial* 1928, n.p.
Barkan, Elazar. 1996. *The Retreat of Scientific Racism: Changing Concepts of Race in Britain and the US Between the World Wars.* Cambridge: Cambridge University Press.

Bonfil Batalla, Guillermo. 1996. *Mexico Profundo: Reclaiming Civilization*. Austin: University of Texas Press.

Cornejo Bouroncle, Jorge. 1949. *Sangre Andina: Diez mujeres Cuzqueñas*, Cuzco: H. G. Rozas.

Corominas, Joan. 1965. *Diccionario crítico etimológico Castellano e Hispánico*. Madrid: Editorial Gredos.

de la Cadena, Marisol. 2000. *Indigenous Mestizos. The Politics of Race and Culture in Cuzco, Peru, 1919–1991*. Durham: Duke University Press.

———. 2001. "Reconstructing race: Racism, culture and mestizaje in Latin America." North American Congress on Latin America (NACLA) Report on the Americas. 34(6) 2001: 16–23.

Escalante, Jose Angel. 1910. *Apuntes acerca del problema de la immigración Europea en le Peru*. Tesis, Archivo Departamental del Cuzco, Fondo Universidad Nacional San Antonio Abad del Cuzco. Libro 9-G.

Findlay, Eileen. 1999. *Imposing Decency: The Politics of Sexuality and Race in Puerto Rico*. Durham: Duke University Press.

Flores, Timoteo. 1918. *Estudio Sicológico del Sentimiento Indígena*. Unpublished thesis, Universidad del Cuzco.

Friedlander, Judith. 1975. *Being Indian in Juayepán: A Study of Forced Identity in Contemporary Mexico*. New York: St. Martin's Press.

García, Uriel. 1937 [1930]. *El Nuevo Indio*. Cuzco: H. G. Rozas, Sues.

———. 1949 *Paisajes Sud-Peruanos*. Lima: Cultura Antártica.

Gill, Lesley. 1994. *Precarious Dependencies: Gender, Class, and Domestic Service in Bolivia*. New York: Columbia University Press.

Gilroy, Paul. 2000. *Against Race. Imagining Political Culture Beyond the Color Line*. Cambridge, M.A.: Harvard University Press.

Guy, Donna. 1991. *Sex and Danger in Buenos Aires: Prostitution, Family and Nation in Argentina*. Lincoln: University of Nebraska Press.

Lira, Jorge A. 1960. "La mujer andina y la biblia." *Revista del Instituto Americano de Arte* (10): 111–117.

Mariscal, Alejandro. 1922. *La Delincuencia Precoz y su Penalidad*. Unpublished thesis. Universidad del Cuzco.

Masis, Horacio. 1927. "Las Walkirias del Ande." *La Sierra* 2(10): 15.

Mintz, Sydney. 1985. *Sweetness and Power: The Place of Sugar in Modern History*. New York: Viking Press.

Nieto, Luis. 1942. "Romance de la Barbaracha." *Revista del Instituto Americano de Arte* 1(1): 25.

Poole, Deborah, ed. 1994. *Unruly Order: Violence, Power and Cultural Identity in the High Provinces of Southern Peru*. Boulder: Westview Press.

Rénique, José Luis. 1991. *Los Sueños de la Sierra: Cusco en el Siglo XX*. Lima: Cepes.

Sotomayor Pérez, José. 1984 [Cuzco 1958]. *Análisis Testimonial de un Movimiento Urbano*. Cuzco: CERA Las Casas.

Stepan, Nancy Leys. 1982. *The Idea of Race in Science: Great Britain 1800–1960*. Hamden: Anchor Books.

Stoler, Ann. 1994. *Race and the Education of Desire.* Durham: Duke University Press.

Valcárcel, Luis E. 1978 [1928]. *El Mito de Kori Ojllo. In Tempestad en los Andes.* Lima: Editorial Universo.

———. 1981. *Memorias.* Lima: Instituto de Estudios Peruanos.

Young, Robert. 1995. *Colonial Desire: Hybridity in Theory, Culture and Race.* London: Routledge.

3. Gender in Movement(s) ↪

8. Engendering Leadership ✐

Indigenous Women Leaders in the Ecuadorian Andes

Emma Cervone
(Translated by Emma Cervone and Deborah Cohen)

INTRODUCTION

We women also want to achieve those positions that our male companions achieve, so we can at least sit in good chairs and also move forward, even though we are women, even though we are indigenous. (Juana 1998)

With this statement Juana, an indigenous leader from the Ecuadorian highlands, synthesizes the formation of an indigenous female leadership, a recent phenomenon breaking a long history of women's political invisibility. While the development of the indigenous movement led to the emergence of political leaders who represent Indians in diverse spheres of national life, this leadership has been open almost exclusively to men. To understand how indigenous women are questioning their political invisibility I focus on two leaders from the Ecuadorian highlands, María and Juana, by reconstructing their paths to leadership.[1] The emergence of female leaders is not linked to a pre-existing process of politicization of gender consciousness among indigenous women. Both María and Juana became leaders through their own efforts as structural factors led them to assume responsibilities within their communities and organizations. Though mostly still manifested in isolated cases, this coalescing of structural factors with female indigenous agency nonetheless legitimizes female leadership in representing Indians in general and indigenous women specifically, in both local and national arenas.

María, a married and well-educated Quichua woman in her late thirties, was elected mayor of the canton of Sumaq in the Cañar province in 1996, as a result of the political empowerment of the indigenous movement both locally and nationally. Supported by her family in the process of leadership formation, María started out with educational responsibilities as an *alfabetizadora* (literacy instructor), and later became president of the local indigenous organization. In exercising leadership in the public office of mayor, María had to fight against racial and gender prejudices as the mestizo council members and local population constantly tried to boycott her work. In contrast to María, Juana, a married and less-educated Quichua woman in her mid-forties from the province of Chimborazo, was not from a local powerful family and was never supported by her relatives. She had to struggle against illiteracy, domestic violence, and poverty. Structural factors including male migration and the impact of a policy promoting gender equality sponsored by national NGOs and international aid agencies helped Juana rise to prominence as a leader. She became president of the cantonal Organization of Indigenous Women of Llacta in 1997 and representative of women in the local indigenous parliament.

While the factors pushing María and Juana to prominence differ greatly, the two cases illustrate how indigenous women can redefine the conditions that undergird their subordination. These conditions are legitimized by the notion of public and private spheres introduced to indigenous politics through the modernizing discourse of the Ecuadorian state. This public/private dichotomy defines gender roles in terms that reinforce the exclusion of indigenous women from positions of leadership: Domesticity, assigned to the private sphere as the feminine realm par excellence, has been interpreted in indigenous cultures as the antithesis of political participation. The latter, ascribed to the public sphere, has become the exclusive domain of indigenous men. Women like María and Juana are redefining this dichotomy by reinterpreting the domestic as a sphere for the production of social values. Their cases illustrate how particular political and social processes lead to the redefinition of gender roles and identities within a given community (cf. Ortiz, this volume) even as they challenge the assumption that women who become leaders fight openly against gender discrimination.

In Ecuador—as in many Latin American countries where social hierarchy is informed by colonial racial prejudices—gender cannot be separated from race and ethnicity. Indigenous women's struggles take place in a context of highly politicized ethnic mobilization in which Indians fight for the participation of indigenous nationalities in the political, cultural, and social life of the country. Since the mid-twentieth century, processes of modernization and democratization functioned as a catalyst toward the politicization of ethnic identity and the consolidation of indigenous movements initiated in the late 1920s. This led to the emergence of an indigenous intelligentsia whose political discourse recalls the abuses and violence suffered by native peoples in order to justify the contemporary indigenous struggle against their white-mestizo[2] exploiters (cf. Pallares 1997). The participation of Indians in the modernizing project of the state comes to symbolize the indigenous vindication from their alleged primitivism as well as their capacity to resist the homogenizing policies of the modern state. In this context, all the accomplishments of the indigenous struggle acquire a symbolic efficacy as not only responses to their material needs as peasants, but also indisputable evidence against the discriminatory imagery constructed by the white-mestizo elite around the Indians for five hundred years.

In this context of ethnic claims, Indians portray indigenous cultures and societies as antithetical to the social injustices and abuses of what they call Western societies. Indigenous social relations, moral values, and cultural manifestations become the ideological foundations for opposing class, racial, and gender inequalities. In this ideological discourse of equality, the supposed gender complementarity of indigenous societies differs profoundly from those forms of gender discrimination identified with "Western" societies.

This highly politicized discourse of gender equality conceals forms of discrimination affecting indigenous women such that discussions of gender issues become highly politically sensitive. This forces women to produce distinct forms of leadership, the discourse of which combines ethnic and gender consciousness feasible in the spaces in which they exercise their leadership. Even so, in their political discourses, female leaders do contribute to the politicization of gender identity by starting to question the construction of gender relations within their own communities, and doing so within a broader commitment to indigenous people's ethnic struggles.

Forms and Spaces of Subordination

As studies of gender relations in the Andes have shown, the sexual division of labor regulating certain aspects of domestic and community life cannot be regarded as an indicator of gender equality[3] in realms related to the market economy and communal decision-making processes. In neither realm are there equal opportunities or access, since the construction of gender identities defines distinct forms of participation and spheres of work. Specifically, in both Sumaq and Llacta, women's participation in community politics is indirect, almost invisible, while men may publicly express points of view that women express in other contexts; and men may speak and make decisions in community meetings. The same division of labor occurs in organizational meetings, where women's active participation is still modest compared to that of men. As Olivia Harris (1986) has shown for the Laymi of highland Bolivia, gender complementarity among Ecuadorian Quichua Indians relies on asymmetric relationships between women and men, far from synonymous with gender equality.

Both the construction of gender relations within contemporary Quichua societies in Ecuador and women's invisibility in the political sphere are linked to the network of power relations between Indians, civil society, and the state. The gendered symmetry between the androcentric and patriarchal construction of the national power structure, on the one hand, and the actual political control exerted by indigenous men within indigenous societies, on the other, has been further reinforced with the "modernization" of the Ecuadorian state and economy. Since the mid-twentieth century, modernization has provoked substantial changes in the system of values legitimized by Quichua societies in the construction of leadership. Although traditional forms of social and political prestige (such as the cargo system) still play a key role in the construction of leadership, formal education, experience in interethnic negotiations, and, most recently, participation in development projects, have all become pivotal criteria for community affirmation of local as well as national leaders (see Cervone 1997). These new attributes required of a "modern" leader have further reinforced the exclusion of indigenous women from political participation as women have had less access to formal education than men,[4] are often illiterate, do not have good command of Spanish, and consequently lack the basic qualifications necessary to gain access to political responsibilities that often involve the white-mestizo world. Most of the women we interviewed for the project considered a lack of formal education to be the primary factor limiting their direct political participation. As one woman put it, "My biggest fear is to talk in meetings because they say that women know nothing" (oral testimony, Sumaq 1997). This woman's testimony eloquently illustrates the extremely negative impact that women's lack of formal education has had on the construction of female indigenous identity. Women's illiteracy is transformed into a condition of social and political inadequacy that silences them in public—even when the meetings are held in their own vernacular language in their own communities. This assumption of inadequacy becomes a constitutive part of a female identity ("women know nothing") that defines women as socially inferior to men. As Harvey (1991) remarks, indigenous women in the Andean region can exercise political pressure but they never have the last word.

Quichua women's lack of self-confidence can be explained by interrelating ethnicity, gender, and place in ways that move us beyond thinking of the dichotomy public/private as describing a simple spatial distinction and instead show it to be constitutive of political spaces characterized by complex social relations and their meanings. In the case I analyze here, "public" and "private" are framed within a context of interethnic conflict saturated with discriminatory images against Indians. For instance, in the modernizing discourse of the Ecuadorian state Indians, with their traditional social and economic settlements, have been considered obstacles to the full modernization of agricultural productivity and are therefore deemed responsible for the nation's economic backwardness. This "indianization" of poverty has been built around a spatial dichotomy in which the city—the place of the white-mestizos—is a synonym for modernity and is opposed to the community—the place of the Indians, the cradle of tradition, and, therefore, antimodern (see Ibarra 1992).

Within Quichua societies in Ecuador, this modernist discourse of racialized spatial oppositions affected women in particular. In this context, "public" no longer refers to a specific place, but rather to a space where Indians have to negotiate with white-mestizo society. The "private" sphere, by contrast, has come to coincide with the domestic sphere, itself linked to those productive and reproductive activities that have remained under women's jurisdiction because they are tied to the subsistence of the household.[5] With the increasing market dependency of the indigenous economy, men have made incursions in modern spaces such as cities, especially through temporary migration. Meanwhile women, due to their direct relation to the community space, have come to stand for and be associated with tradition and all the negative connotations attached to it.[6] As a result, women's access to public activities, that is, activities that imply interethnic interactions, continues to be very limited and this sphere, defined in terms of certain activities related to politics and the market, has been gradually assimilated as a male domain.

These spatial constructions have led to the definition of yet another gendered spatial dichotomy that emerged from the structural conditions to which the indigenous economic and political systems have had to adjust ever since the agrarian reform of the 1960s. These conditions led to the creation of the opposition between "community" and "home," in which the former becomes an interstitial space linking the indigenous world to the outside, that is, to the world of the white-mestizos. In fact, community leaders have to deal with NGOs, state representatives, and various institutions that may be relevant to their communities' development. Conversely, home becomes a purely domestic space, the only indigenous sphere free from direct intrusion by the outside world.

The intertwining of the dichotomies city versus community and community versus home has led to a division of labor that excludes women from market-related activities both inside and outside the community. Women's exclusion often encompasses extracommunity activities that do not necessarily imply interethnic interactions, but simply exceed domestic activities, evidenced by the fact that men do not allow their wives to work far from home or participate in activities that would require them to stay away from home. Domestic violence then becomes a way of restricting women's access to the public sphere (cf. Massey 1994, 179). "My desire is to be able to return home without being afraid of my husband," commented one of the women (Sumaq 1997), suggesting that it is difficult for women to leave their home and community by themselves without provoking a domestic fight.

The articulation of these dichotomous oppositions is linked to the realm of interethnic conflict that excludes Indians from the political and economic national project. However, the subordination of women derived from this articulation of spatial dichotomies has been redefined by structural changes occurring within indigenous society and economy since the 1960s. As Juana's case shows, the increasingly common phenomenon of male temporary migration to the city has led women to assume responsibilities within the community as they find themselves replacing their absent husbands. In this transition they have come to play a crucial role in the political and material survival of the community. It is precisely this structural shift that has allowed some indigenous women like Juana to become visible and start to exercise leadership.

EXEMPLARY HISTORIES

Conquering the Town

The cases of María and Juana differ significantly in terms of the structural factors that helped them become leaders. The distinct social conditions and circumstances affecting the two women and their rise to prominence have, in turn, influenced the spaces in which they exercise their leadership. While María assumed the office of mayor in a majority mestizo district, Juana dedicated her soul and work to her native community and to cantonal, or what is referred to as second-level indigenous organizations (hereafter SGOs) in her district. Eventually Juana established the district's first women's SGO. The different experiences and challenges that each woman has had to confront also contributed to the production of distinct political discourses. The strategic approach of each discourse is, in turn, directly related to the specific space in which each leader acts.

In María's case, her travels and experience together created the necessary conditions for shaping a leadership that is rarely available to Quichua women. This exceptionality resulted from the coexistence of several factors that permitted María to continue building her leadership and skills. First, María came from a prestigious family with a tradition of leadership at the local level. Her family's wealth also meant that she and her siblings all received formal education. As María remembers, her mother always pushed her to study and work. In contrast to many indigenous women, María finished high school, acquiring a status that allowed her to be chosen by her community at various times for tasks that required formal education. For instance, she worked as treasurer in a housing cooperative and, after that, as literacy teacher in a program in the Roldós-Hurtado government, the first government after the country's shift to democracy in the 1980s. After getting married, however, María could not have continued working and raising her children without the help of her mother or sisters.

María's position as Provincial Coordinator of Early Childhood Education (*Educación Parvularia*) was another factor that contributed to her rise to prominence because it gave her the opportunity to begin negotiating and engaging with the white-mestizo world. "In offices they laughed, they thought that I wasn't capable of heading up an institution, simply because I am an indigenous person or because I'm a woman" (María 1998). These educational responsibilities put María in contact with many indigenous communities and allowed her to become known among her own people. This exposure gave her the opportunity to fill an important post rarely assumed by a woman: the presidency of a second-level organization, the Peasant and Indigenous Association (AINCA) of Sumaq, which she directed for two years.

María had to exercise leadership in the face of poverty, organizational difficulties, and bureaucracy; but the responsibility of being president allowed her to claim a commitment to and experience in the public sector as well as in administration. The development projects entailed handling and being in charge of large sums of money, as well as the responsibility of helping affiliated communities move forward. Because of this experience, María was well situated to take on the position of spokesperson on the District Board, her entrée into local mestizo society. Manuel, her husband, a recognized leader of the indigenous sector and founder of AINCA, provided the link. María's role as wife of an important and experienced regional leader guaranteed her husband's help in these matters. Nonetheless, they had to negotiate as a couple because her activities took her out of the house, sometimes requiring her to spend several days away. "There are always problems but you have to talk about them and try and work them out. ... If it's that the husband doesn't want to understand, then you have to do whatever is possible so that he comes to a meeting or a workshop. ... Because there he'll see that the woman is making a sacrifice, too, she's working" (María 1998).

María's election as mayor in the 1996 national elections did not stem exclusively from her experience and dedication. Her nomination and final victory were also facilitated by the opening granted the indigenous sector by progressive social forces. These forces, which were not aligned with any political party, coalesced around the Pachakutic Movement to pursue social changes for the most marginalized sectors of Ecuadorian society. This political opening allowed indigenous people to occupy many enclaves of local power, especially those in districts of the highlands and the Amazon, where a majority of the local population was indigenous.

The decision to nominate María as mayor of Sumaq resulted from negotiations between the AINCA and the Pachakutik Movement, the latter an organization that had previously only offered council member positions to Indians. Initially the organization's assembly decided to nominate María's husband, the director and founder of the organization. The choice of María came afterward and for strategic reasons. At that time María had been re-elected president of AINCA. Her election as mayor would have represented a victory for the whole organization and, ultimately, for the indigenous sector, with the symbolic conquest of the mayor's office. Moreover, choosing her would have symbolized taking an independent position vis-à-vis the political parties that also had proposed her husband as candidate. In that context María turned out to be the ideal candidate: She had amassed leadership experience, she had the support of a husband who was a recognized leader, and she was an indigenous woman. She also represented an indigenous organization and not the will of political parties. "The parties wanted to launch Manuel, my husband, as the candidate, and me also as a council member. I didn't accept, because I wouldn't betray my people. I have to say, the indigenous organizations have more rights, they are working for indigenous people, so they believed it was necessary that indigenous people assume power" (María 1998).

"We Also Can"

Juana's experience, by contrast, is an eloquent example of how discriminatory gender roles and identities impede indigenous women from gaining access to political power and often condemn them to a life of abuse and violence. Coming from a poor family that was not part of the local ethnic elite, Juana never had her family's support for her formal schooling. "When I was young, I didn't like [to study], to tell the truth, I didn't like to study. ... More than anything, it was because they said in earlier times that women don't need to know how to read and write. ... my father heard this and didn't let me finish primary school" (Juana 1998). Juana sought relief in marriage, yet, in a scenario common to many highland indigenous women, she did not receive better treatment from her husband. Juana's family did not support or defend her against her husband's attacks of jealousy and violence. Her own father told her: "I don't want to know anything. If your husband comes, I'll turn you over to him. ... Married, that's it. So he may kill you, but you have to die in the hands of your husband" (Juana 1998).

The factors that helped Juana build her leadership role are linked to the structural shift I referred to earlier. Although the modernizing of Ecuadorian economic and political structures had negative impacts on indigenous women, factors such as male migration provided an unexpected opportunity for Juana to move into a position of leadership, while also leading her to question the discriminatory gender practices and images of which she had been a victim. When Juana's husband, like many other men in her community, started to migrate to the city on a periodic basis, Juana had to replace him in his positions as community treasurer and as treasurer of an irrigation system association incorporating seven communities. Another structural factor crucial to Juana's future success as a leader was the impact of a new regional development policy that sought to fight gender discrimination. Through local NGOs, various international entities sponsored small-scale productive projects aimed at improving women's social and economic position. One of these was the Rural Women and Families Development (ALA) Project financed by the European community in conjunction with the Ministry of Agriculture. Carried out in the province of Chimborazo, the ALA Project implemented productive microactivities as economic support for indigenous women. The project's principal objective, however, was to raise consciousness among women around issues of gender equality and to foster women's empowerment. One of the requirements for obtaining a loan for such productive microactivities was the founding of women's groups at the community level. Through training and consciousness-raising courses Juana received from the ALA Project, she overcame her educational deficiencies and insecurities and learned to trust herself and her capabilities: "'Don't be afraid,' they told me. 'You have control over seven communities, so why should you be scared? Be strong, some day you'll be a leader, don't be afraid, participate, say whatever you want'" (Juana 1998).

Juana's participation in the ALA Project initiated her transformation into a leader, as she gradually acquired more prestige within the local Union of Peasant Indigenous Organizations, or UOCIG. Her first leadership position was president of the ALA Project in Llacta, after which she was nominated women's secretary of UOCIG, a position she held until she was elected to the indigenous parliament of Llacta as representative of indigenous women. In addition, Juana was active in creating the Organization of Indigenous Women of Llacta, of which she also became the president.

Initially, Juana's involvement and leadership in the UOCIG created difficulties with her husband. Her extracommunity activities sparked his jealousy and, later, physical aggression. But her talent for the work and her road to leadership helped demonstrate her capabilities and dedication to him, her family, and her community. As Juana explained, using the discourse of the ALA Project, "My husband now acts like an adult man, he finally respects me. Besides, he also participates in courses. Now we understand and treat each other as equals. So, if for some reason I am not here, he does the things that I must do, or when he's not here, I help do his things" (Juana 1998).

Once Juana began to be politically active, she also gained the respect of other members of her husband's family. For example, her sister-in-law, who used to mistreat her, began accompanying her to meetings and courses and helped look after the young children.

Through her participation in the ALA Project, Juana gradually developed her own perspective on the gender discrimination that she endured by linking it to a wider context of ethnic discrimination. "In the old days we women were discarded, left to one side, because there's a lot of machismo. Male companions have always been trampling on us, we were underneath their shoes. That was what they learned from so many past years because the *hacendados* [landowners] had us under their shoes. For that reason there are still male companions today who don't give women priority" (Juana 1997).

Juana's testimony is a clear articulation of the gendered symmetry between indigenous and nonindigenous forms of male political domination that women leaders like Juana perceive to be interdependent. This articulation allows indigenous women to combine ethnic and gender consciousness and project their struggle toward wider contexts of discrimination that involve not only their men but also the white-mestizo society.

CHALLENGES AND RESPONSES

Redefining Domesticity

The distinct factors helping María and Juana achieve their respective positions presented each woman with different challenges for the exercise of leadership. Juana's case shows that in the community sphere women have to confront the realities of gender relations and fight against those practices and values that legitimize violence again women and render their participation in community politics invisible. The political rhetoric of grassroots women leaders such as Juana is constructed around these challenges. Although women's leadership was not initially linked to the politicization of gender consciousness, indigenous women have profited from the structural openings that fostered its emergence in order to attack various forms of gender discrimination. In Juana's discourse and political practice, we can see the extent to which she incorporated gender into her perspective and analysis. A consciousness both of women's capabilities and of gender as a set of power relations comes to constitute a fundamental framework of her political understanding. "So the project might end, but I never want to leave my women companions because they might return to where they were before. I want them to continue growing, learning. ... Male leaders ... are more prepared. Women still lack this, we must prepare ourselves ... " (Juana 1998).

Female grassroots leaders like Juana do not hide the existence of domestic violence and the inequality of rights between men and women; rather, they denounce it even as they claim that it is changing.[7] Juana never denied the reality of the domestic violence she had experienced. Another grassroots leader from Sumaq said: "We women have always suffered. As a girl, I remember that my father hit my mom for no reason. My first husband also hit me a lot. We weren't worth anything. They didn't let us leave the house, nor did they send us to school. But now things are changing. In our organization we women talk and they listen to us more" (oral testimony, Sumaq 1997).

In communal political spaces and those connected with grassroots organizations, women leaders promote gender solidarity by bringing their challenges and claims *as women* to organizations to carve out space and political legitimacy for themselves. In this way they are able to influence the decision-making process in the indigenous movement.

The incursion into spaces traditionally denied to women, such as the market and local politics, is not the only strategy that grassroots women leaders adopt in order to defend their right to equality. They also reinterpret gendered spaces and identities that feminize the domestic sphere and limit women's access to spaces of interethnic interaction, as a way of legitimizing their political practice. As the following women's comments indicate, activities strictly related to the reproductive and domestic sphere, such as child care, become distinctive markers of responsibility that extends to women's political participation. As a result, women come to perceive the domestic sphere, not as a private realm, but as a public space where social values are produced.

> We women aren't like men. Without thinking they'll suddenly find themselves ... going out and pounding down drinks and having discussions or arguments. We women, we finish meetings and off [we go] to our houses. Or, suddenly we are amongst ourselves as women and we ask each other, "how's it going for you in the group?" or "how is the president doing?" ... I think that we are providing an example for the men. (Juana 1998)

> We as women, with our sense of responsibility as mothers, can contribute to the design of policies of indigenous organizations, not only for women. (oral testimony, Quito 1997)

> I worked in a childcare center and it was very important. By chance only men know how to earn money? We [women] have also moved the community forward, with the childcare center, and now with the mill project. We [women] also know how to produce. (oral testimony, Sumaq 1997)

These comments also have recourse to a political discourse that distances itself from notions of gender equality as conceived by liberal western feminism, according to which women aim to be considered equal to men. Drawing upon the political discourse of the Ecuadorian indigenous movement, leaders like Juana advocate for "equality in diversity" for the two genders: Women are not just like men, they are different; but it is precisely their difference that legitimizes their capabilities and establishes their rights to the same opportunities as men. At the same time, the pervasiveness of ethnic identity existing in and tied to political practice precludes looking for a solution to gender discrimination separately from mechanisms that address ethnic discrimination. Even as grassroots women leaders fight to break through their invisibility as women, they do not see men as rivals in the struggle. "With male companions, you should work and discuss so that they understand but also, so that we all move forward, with them, all indigenous people together" (oral testimony, Quito 1997).

As Juana remarked, while it is important to create a space in which women can reflect on their situation and problems they face, it is also crucial that women work through their political understanding with their male companions. Only through cooperation, through working together will they be able to transform gender relations.

Redefining Motherhood

The interethnic political sphere to which María gained access presented different challenges than those of the local community sphere in which Juana acquired a leadership position. This interethnic sphere is an arena in which the indigenous population comes face-to-face with both the state and white-mestizo society. To understand the responses and strategies that female leaders like María adopt when they hold public offices, we must contextualize their political practices within a broader framework of ethnic mobilization. In this way we can see them as among the responses to ethnic discrimination put forward by the indigenous movement.

In María's case, dependency on male support and recognition was blatantly manifest. The gendered symmetry between indigenous and non-indigenous male domination, aggravated in Sumaq by interethnic conflict, constantly threatened to delegitimize María's performance as mayor. The local mestizo population boycotted many works that María tried to implement in the town. Given this hostile and conflictive context, María was accompanied in public by her husband and discussed any decision she had to make with him, knowing that people would take her more seriously with her husband present. She also adopted a discourse that claimed Quichua men and women had the same rights and opportunities and should share equally in tasks and responsibilities. By emphasizing gender complementarity, María relegated gender discrimination to the margins. In contrast to Juana, María took this ideological stand to an extreme. She never worked to build networks of support with other indigenous women of her sector; moreover, she considered the issue of gender to be more relevant to the conditions of nonindigenous women, maintaining that indigenous women did not face gender discrimination. "Yes, I'd heard about other leaders in high school but I didn't know them or their way of participating. I've never made contact with other women" (María 1998).

How are we to understand María's responses to her challenges? As I pointed out previously, the discourse of the indigenous movement, when directed toward the white-mestizo interlocutor, juxtaposes an indigenous society grounded in ideals of equality and perfection with the violence and abuse rampant in Western society. The rhetoric of indigenous women leaders operating at the national level emerged from this context of interethnic struggle. Its target is thus not only men, but also the white-mestizo society more generally. Given this context, women must defend both their gender and their race, making indigenous men less responsible for the causes that erase women's political participation. According to some of these women leaders, such as Blanca Chancoso, "indigenous women made themselves invisible as a way of defending themselves from violence and aggression from conquerors and colonizers" (CEPLAES workshop, Quito 1997). This invisibility, remarks Chancoso, does not imply that indigenous men and women do not have the same rights and opportunities in daily life. Rather, indigenous notions of complementarity and the sexual division of labor support an ideology of gender equality within indigenous societies that is counterposed to various forms of gender discrimination within the white-mestizo and mestizo world. This same discourse denies the existence of domestic violence in the indigenous world or justifies it by attributing its origin to exclusively white-mestizo and Western societies.

We must understand this ideology of gender equality, therefore, as a strategy used by women leaders to carve out and defend a space for women within the indigenous movement. In the absence of a women's ethnic movement able to legitimate and defend its own understanding of gender rights, those able to validate women's public leadership and its language are men, specifically male leaders. Female leaders like Blanca Chancoso and María representing the indigenous population vis-à-vis the white-mestizo society, adopt the same generic discourse on Indian-ness produced by male leaders. As María stated: "In the countryside men and women have equal daily activities... everyone shares the same vision of the world... in the urban, mestizo sector the issue is different and indigenous women are assigned different tasks than those given to and of indigenous men" (María 1998).

In the case of women's leadership in national-level offices, therefore, gender-oriented claims operate on a symbolic level. In these political spaces, women challenge the hegemonic linking of women with tradition by disputing its negative connotations. The Indian woman, "born from the womb of mother earth, made fertile in our first nations, building the way, illustrating tomorrow" (from CONAIE 1995), becomes romanticized as the hope for the future. Indigenous women are presented as uniquely able to guarantee the survival and reproduction of indigenous populations. Indigenous traditions, which women both create and protect, are the only guarantee of reproduction and survival of indigenous cultures in the country. "Sisters, ... we are here precisely to reflect on the beauty, those flower gardens of a thousand colors, the beauty of women, the wisdom of mothers and motherhood, which is the reflection of mother nature" (CONAIE 1995, 8).

If, on the one hand, the language of modernity discounts tradition by associating it with the antimodern, on the other hand, a discourse shaped by ethnic claims lays out other relationships and values that imbue the term "tradition" with more positive meanings and associations. Tradition becomes associated with "*lo propio,*" that of one's own. In this way "tradition," that which is indigenous, is presented as the only just and balanced alternative to Western forms of injustice and destruction. Ironically, this essentialist discourse that "naturalizes women's identity by stereotypical identification with Mother Earth" (Muratorio 1998) does not marginalize women's voices. Instead, it is reinterpreted by women within a framework of ethnic struggle as a way of legitimizing their political participation within the movement, especially when women occupy positions in which they represent indigenous nationalities vis-à-vis the rest of society and the state. Once again, female leaders are questioning hegemonic definitions of public/private by rescuing motherhood from the private sphere and attributing to it social values that give women the right to be publicly visible.

CONCLUSIONS

Indigenous women in Ecuador are redefining gender relations and occupying a space of political participation within the indigenous movement. Instead of following the path of gender-specific political mobilization in opposition to men, indigenous female leaders are redefining concepts such as public/private that had established a gendered symmetry between indigenous and nonindigenous forms of male leadership while condemning women to invisibility. The articulation of ethnic and gender dimensions of discrimination is manifest in the political practices that operate both within the realm of local community leadership that Juana occupies and within the realm of national leadership that María occupies. The different ways in which the two leaders respond to their challenges emerge from the different spaces in which they exercise their leadership. While grassroots leaders like Juana challenge the public/private dichotomy operating on the level of practice, national-level leaders like María question the opposition at the level of semantic and symbolic resignification. In the first case, women attribute social values to the private sphere of domesticity in order to prove the legitimacy of female leadership. In the second case, they reinterpret the meaning of tradition and motherhood as elements that exalt the role of women within indigenous cultures and therefore justify their participation in the struggle of the indigenous movement.

Yet, despite differences between local and national political discourses, both grassroots and public women leaders agree on one point: Their work against gender discrimination must be carried out in conjunction with indigenous men as a part of the Quichua people's struggle. In the cases both of grassroots and national-level female leaders, gender relations are redefined through cooperation and women's political legitimacy is linked to a political practice that seeks the well-being of the indigenous population as a whole.

This path to leadership however, is not free from challenges and vulnerabilities that threaten the legitimacy of female leaders. Both María's and

Juana's leaderships are built on structural and strategic factors that, along with their personal strengths, have pushed them to take on political responsibilities vis-à-vis the local population. But the leadership of both women still suffers from a lack of legitimacy. In Juana's case this stems from her background and her political formation; in María's case, it stems from her dependency on her husband's reputation to carry out the demands of her office.

Juana's challenge is also related to gender-based development projects greatly impacting indigenous women's processes of consciousness-raising and politicization. Many of these projects encouraged women to create groups and organizations that, while oriented toward productive microactivities, have as their main goal to foster equality between men and women. Women's leadership tied to those projects runs the risk of being confined to "women's space," thereby separated from the organizational decision-making processes in which men participate. In some cases, the extension of political participation to women within indigenous organizations has responded to a specific logic emerging out of funded projects. For instance, many women who have assumed tasks within indigenous organizations receive posts as directors or secretaries "of women." Juana's case points out the local impact of European economic community projects: Because of these projects and the availability of funds, the organization UOCIG created a women's ministry and offered Juana its directorship. Today's participation by women in development activities and in the political life of organizations seems to be a strategic response to the indigenous movement's needs for economic resources, brought to the community through specifically gendered development projects.

In María's case, her political vulnerability derived from two factors. First, her candidacy resulted from the political strategies of local indigenous organizations, rather than the legitimacy of her leadership per se. Second, she depended heavily on her husband's support. The ideology of gender equality has not been sufficient to legitimize María's leadership, especially since she did not cultivate a network of female solidarity that would have legitimized her, independent of her husband.

Comparing how the two leaders have taken on the challenges of their positions and redefined gender relations in their spheres of action, it is clear that grassroots leaders are more successful in claiming political legitimacy. Gender-oriented practices provided Juana both the recognition and support of other women and established her legitimacy at the local level—despite, but not against, indigenous male leaders. Juana is a member of the local indigenous parliament precisely because she represents women's political force at the local level.

Conversely, the ideology of an existing gender equality discouraged María from questioning and interrogating the ways in which women holding public office are forced to depend on male leaders for their own legitimacy when the former face interethnic conflict. If, within the indigenous movement, this ideology gives women access to a strategic discourse that benefits and enables them to gain local recognition, when women confront interethnic frictions, it does not bring the same political efficacy. In the interethnic context, the political presence of indigenous women can turn into a circumstantial consideration indicating structural change. In addition, omitting or marginalizing the scope of gender claims often creates tensions between women holding national-level offices and grassroots leaders, whose political practices incorporate women's perspectives and claims.

Still, the path appears long and tortuous. Existing challenges for indigenous women (structural changes, various forms of gender and ethnic discrimination, politicization of ethnicity) yield to and are framed within a network of power relations in which gender claims have not yet acquired political legitimacy, either in relation to the state and civil society, or within the indigenous movement itself. Yet indigenous women are fighting to redefine the conditions that have caused their exclusion and invisibility. They exemplify the possibility of struggling against various forms of gender discrimination without necessarily rejecting "traditional" gender roles and values by giving those roles and values new social and political meanings.

NOTES

1. The chapter is based on a research project, "Indigenous Female Leaders: Lessons and Challenges," carried out by the Ecuadorian NGO Centro de Planificación y Estudios Sociales (CEPLAES). Names of the leaders and their respective towns are pseudonyms.
2. This term was coined by social scientists to refer to the Ecuadorian elite who consider themselves white. The term differentiates this elite from people (usually of the popular classes) who consider themselves mestizos.
3. See Hamilton 1998; Harris 1986; Harvey 1991; Weismantel 1988.
4. According to several oral testimonies, many women in their thirties and forties did not receive formal education when they were children, while their brothers were sent to school because it was believed that men needed to know how to read and write. This differentiation was naturalized by the belief that men learned more easily than women. For a different gendering of local educational ideologies see Hurtig, Chapter 1, this volume.
5. By "domestic" I am referring to the sphere that involves household productive and reproductive activities.
6. For a comparative case on Peru see de la Cadena 1991.

7. These conclusions are based on the testimonies of female grassroots leaders from different cantons interviewed during the research project. They all mention domestic violence as one of the reasons why it is difficult for women to pursue their political aspirations.

REFERENCES

Centro de Planificación y Estudios Sociales (CEPLAES). 1998. *Mujeres contracorriente. Voces de líderes indígenas.* Quito: CEPLAES-FEG.

Cervone, Emma. 1997. "The Return of Atahualpa. Ethnic Conflict and Indigenous Movement in the Ecuadorian Andes." Ph.D. diss., University of St. Andrews, Scotland, UK.

Confederación de Nacionalidades Indígenas del Ecuador (CONAIE). 1989. *Las Nacionalidades Indígenas en el Ecuador. Nuestro proceso organizativo.* Quito: Abya-yala-Tincui.

———. 1995. *Memorias. Encuentro de Mujeres Indígenas de las Primeras Naciones del Continente.* Quito: CONAIE.

De la Cadena, Marisol. 1991. "'Las mujeres son más indias.' Etnicidad y género en la comunidad del Cusco." *Revista Andina* No. 1: 7–29.

———. 1996. "The Political Tensions of Representation: Intellectuals and Mestizas in Cuzco." *Journal of Latin American Anthropology* 2(1): 112–147.

Hamilton, Sarah. 1998. *The Two-Headed Household: Gender and Agricultural Development in the Ecuadorian Andes.* Pittsburgh: University of Pittsburgh Press.

Harris, Olivia. 1985. "Ecological Duality and the Role of the Center: Northern Potosí." In *Andean Ecology and Civilization: An Interdisciplinary Perspective on Andean Ecological Complementarity,* edited by Shozo Masuda, Izumi Shimada, and Craig Morris. Tokyo: Tokyo University Press, 311–335.

———. 1986. "Complementarity and Conflict: An Andean View of Women and Men." In *Sex and Age as Principles of Social Differentiation,* edited by Joan de la Fontaine. ASA 17. London: Routledge, 21–39.

Harvey, Penelope. 1991. "Mujeres que no hablan castellano. Género, poder y bilingüismo en un pueblo andino." *Allpanchis* 38: 227–260.

Ibarra, Hernán. 1992. *Indios y cholos. Orígenes de la clase trabajadora ecuatoriana.* Quito: El Conejo.

Massey, Doreen. 1994. *Space, Place, and Gender.* Minneapolis: University of Minnesota Press.

Muratorio, Blanca. 1998. "Indigenous Women's Identities and the Politics of Cultural Reproduction in the Ecuadorian Amazon." *American Anthropologist* 100(2): 409–420.

Pallares, Amalia. 1997. "Seeking Respeto: Racial Consciousness and Indian Politics in Ecuador." Paper presented at the nineteenth LASA Congress, Guadalajara, México. Mimeo.

Weismantel, Mary. 1988. *Food, Gender, and Poverty in the Ecuadorian Andes.* Philadelphia: University of Pennsylvania Press.

9. Latinas on the Border ✑

The Common Ground of Economic Displacements and Breakthroughs

Victor M. Ortiz

PROCLAMATION I: A MOURNFUL CELEBRATION

On April 19, 1990, celebrating the bountiful economic benefits of the *maquiladora* industry for the local business community, the mayor of El Paso proclaimed the city to be the "Maquila Capital of the World" (Baake 1990). "*Maquila*" is the colloquial term for maquiladoras, export-oriented assembly line plants that companies from the United States, Japan, and other industrialized countries have established in Mexico. These plants, located primarily along Mexico's northern border region, take advantage of comparatively lower wages and other incentives offered by the Mexican government. The mayor's proclamation referred to the high concentration of these production plants in Ciudad Juarez, the Mexican city immediately across the border from El Paso.

The mayor's personal experience directly attested to these benefits: Her husband owned a maquiladora plant and she was an active businesswoman of long-standing success in El Paso. The pronouncement was made at the inauguration of "Maquila: La Vista Grande" (Maquila: The Big Picture), a week-long series of events that aimed to "increase the level of awareness of the maquila industry and its impact on the local community" (*El Paso Herald Post* 1990). The event was sponsored by the "Twin Plant Wives Association" with the initiative and help of Ms. Marcela Torres,[1] a young entrepreneur involved in the promotion of the maquiladora industry.

The event and the mayor's statement provoked polarized responses, echoing the polemics that have surrounded the maquiladora industry throughout its 36 years of existence. Advocates of the industry stress the economic benefits of employment and business opportunities (Stoddard 1987, *Twin Plant News* 1990) while critics of the industry point to displaced workers in the United States and low wages, limited training, lack of job security, unsafe working conditions, and instances of sexual exploitation for maquiladora workers in Mexico (Fernández-Kelly 1983; Tiano 1994; Peña 1997). The event and its responses also reflected the contrasting impacts that processes of global integration have had for different local groups.

This contrast contributed to the sharp polarization between a growing number of displaced garment workers (most of whom are Mexican and Mexican American women) whose jobs were relocated to Mexico and other third world countries, and another group of local women making unprecedented breakthroughs in business and politics. In other words, while El Paso was on its way to becoming the city with the largest number of North Atlantic Free Trade Agreement (NAFTA)-related applications for layoff benefits in the entire United States,[2] women had attained prominent public positions including a county judge, two state senators, a city mayor, a university president, three city councilwomen, a community college president, and two presidents of chambers of commerce. With the exception of the city council posts, none of these public positions had been previously held by women. Notably, five of the eleven women are Latinas, including the county judge. Latina businesswomen have also attracted public attention due to their extensive participation in local affairs. One such woman is Marcela Torres, organizer of the "Maquila: La Vista Grande" event. In 1992 she received a regional award from the Small Business Administration for her advocacy of minority entrepreneurs. Months later, she appeared in the front cover features of two business magazines with national circulation.

This chapter contrasts the public involvement of two Latina women—one an entrepreneur, the other a labor advocate—who pursued contrasting yet inevitably related paths in the frontiers that global economic integration opened for El Paso. Marcela Torres, the organizer of "La Vista Grande" event, is a small-business owner promoting a shift in the local economy from light manufacturing to professional and technical services. Carmen Rocha, a labor activist, promoted the upgrading of local manufacturing capabilities to maintain the competitiveness of the city in an increasingly globalized garment industry. The two women's paths were related literally and figuratively by a common ground: Both women are natives and residents of El Paso and share a strong commitment to the city, and they both aimed to redirect the economic orientation of the city in response to the ongoing impacts of capitalist reconfiguration. The involvement of the two women reflects their situations as highly skilled women in a particular geographical setting shaped by its intermediary role in the flows and exchanges between Mexico and the United States. At the same time, their shared concerns for local conditions were polarized by a complex tension derived from their ideological and professional positions. The impacts of global economic integration influence these women's options that, in turn, affect how these women influence these impacts. The relationship between Torres's ostensible success and the worsening conditions for the thousands of displaced garment workers poses important questions. Does her class mobility override gender inequalities, or does her success as a Latina show that gender and ethnic barriers have decreased while class inequalities have deepened—as Rocha's mobilization efforts suggest?

PROCLAMATION II: THE WOMAN QUESTION HAS BEEN ANSWERED

With similar complexity to the Mayor's pronouncement of El Paso as the Maquila Capital of the World, author Germaine Greer declared the woman question answered in her recent book, *The Whole Woman* (1999). Tackling perplexing transformations in gender relations in the last three decades, Greer argues that while women have made evident breakthroughs, the salience of many of these breakthroughs belies persistent inequalities. She points out that Western societies have become more masculine, despite the greater participation of women in positions of public influence. Unlike the mayor, however, Greer's proclamation is mostly rhetorical: "The woman question is answered. ... Feminism has served its purpose and should now eff off. ... We all agree that women should have equal pay for equal work, be equal before the law, do no more housework than men do, spend no more time with children than men do—or do we?" (1999, 9).

Like the mayor of the city of El Paso, Greer suggests new scenarios and impending opportunities for previously marginalized groups. However, unlike the mayor, Greer stresses the baffling character of the changes, in which inequality insidiously persists. She points out that breakthroughs remain mostly individual, not having consolidated at the institutional level: "Women may enter political institutions only after those institutions have formed them in the institutional mould" (16). Women's increased presence in the political sphere ironically undermines the representation of women's concerns: "the more female politicians a parliament may boast, the less likely it is to address women's issues" (16). This lack of institutional change fosters a new type of insidious oppression in which individual gains pre-empt collective political advances as a mirage of equality distracts from the original demands. As such, strides toward gender equality have not only slowed in Western societies, but also the insidious mirage of gender equality has spread to all corners of the world with globalization.

Diane Elson and Ruth Pearson expressed similar concerns almost twenty years ago in their examination of the gender dynamics of the "new international division of labor." Looking at employment practices of plants relocated from industrialized to industrializing countries, such as the maquiladora plants in Mexico, they argued that companies not only exploited local gender inequalities but transformed them, mostly for the worse, as an increasing number of women experienced complex vulnerabilities in their formal integration to global labor markets (1984, 29). The authors critically examined the deeply ingrained hope among certain left intellectuals that the "proletarianization" of women would foster gender equality.[3]

Elson and Pearson, like Greer, remind us that the development of capitalism and the participation of women in the labor force have turned out to be more complex than intellectuals, inspired by Frederick Engels, foresaw. For Engels, the incorporation of women into the labor market as wage earners would provide them with resources and options to overcome their subordination. This "proletarianization" alternative was, for orthodox Marxists, the response to the women question. Elson and Pearson noted that the expansion of manufacturing operations from industrialized to industrializing countries in the 1970s and 1980s brought about new forms of gender subordination while partially weakening or reconfiguring others. Their analysis follows the feminist call for an epistemological reformulation that moves beyond reductive class analyses and toward recognition of the dynamic social complexity of ever-changing economic arrangements. This call is heeded by Greer who examines the striking yet limited transformation of gender relations in recent years. The call is central to the conceptual thrust of this volume's aim to desalambrar analytical approaches by taking into account the dimension of place in the transformations and struggles of gendered practices.

The case of El Paso calls into question the linear logic of modernizing assumptions that gained credibility as large numbers of women were incorporated into nationwide manufacturing arrangements since the early 1960s. A central feature of the globalization of production is the importance that women's labor has played, as most of the initial activities relocated were operations typically done by women in the electronics and garment industries

(Fernández-Kelly 1983). In the relentless search for lower costs, businesses and corporations continue to generate and exploit labor pools within their own countries and abroad through the manipulation of dramatic disparities.

This "rationalization" of economic factors results in regional disparities within countries that challenge us to think beyond monolithic constructions or linear developments (Massey 1994). At the opposite end of plant relocations from the first world to the third world discussed by Elson and Pearson, my examination of El Paso focuses on the first world as an instance of deindustrialization in the United States. El Paso is a city in an industrialized country that has been severely affected by the relocation of plants, not into but away from it, thus calling into question the industrial/industrializing divide of north and south. Moreover, the disparities are far from monolithic, but are conditioned by regional, gender, and class particularities. Within each region or site, economic transformations affect local constituencies differently. Polarizing effects within the given constituencies themselves are striking. For example, in El Paso, some women have enjoyed unprecedented opportunities while others were laid-off. These concurrent developments, however, elude an easy zero sum analysis, precisely because of their fragmented and dynamic character. While some jobs were relocated from El Paso to Mexico and Central America, other jobs were simultaneously relocated to the city. Yet, the new jobs seldom employ the laid off workers, whose years of job experience now make them "unfit" for the new positions. Overspecialized in obsolete manufacturing operations, middle-aged workers are seen as too old and practically "untrainable" for the new, better paying jobs. Thus, displaced workers find their options limited to dead-end jobs and public assistance.

At the same time, a number of women made unprecedented inroads in entrepreneurial venues. These breakthroughs are particularly evident in the number of Latina-owned companies in the United States, which tripled from 1987 to 1996, reaching a total of 659,000. This 209 percent increase dwarfs the 47 percent general national rate of growth in new businesses for the same years. While Latina-owned firms only account for one-third of Latino-owned businesses, the progress of Latinas as a group is not negligible—with a 534 percent growth rate in sales and 48.7 percent growth rate in employment (Russel 1998, 33). Still, these growth figures are neither uneventful nor assured. The image of success can be tenuous and very partial. As we shall see, Torres's case suggests that, despite her relative good fortune in the new economic landscape compared to that of the displaced workers, she has not been exempt from exploitation or undue demands based on gender expectations. In this sense, her case manifests the reconstitution of gender inequality as conceptualized by Elson and Pearson and poignantly articulated by Greer.

The sharp polarization between winners and losers is connected. The enhanced mobility of capital in the new arrangement opens almost as quickly as it closes economic opportunities to localities. This volatility imposes constant challenges for small businesses striving to consolidate their markets in innovative and rapidly changing fields. These unstable markets not only affect individual entrepreneurs, they also present a challenge to local policy makers and cities as a whole. Factors such as plant closings and the rapid obsolescence of certain technical skills produce various dislocations for local settings. Some of these dislocations are primarily social, ranging from higher indices of alcoholism to domestic violence to suicide. Others are more technical in nature, as in the retraining of workers and the attraction and retention of new industries. In the meantime, local communities contend with these challenges in the face of dwindling fiscal resources, as their tax bases lose some of their principal contributors in the form of major employers, and the chance of bankruptcy among smaller businesses rises (Perruci and Targ 1988). These dislocations are quite tangible in the city of El Paso. The following section provides a brief overview of the economic history of the city, as it relates to the conditions addressed by the contrasting initiatives promoted by Torres, the local entrepreneur, and Rocha, the labor activist.

EL PASO AS A GLOBAL PLACE

El Paso became a major center for basic sewing operations in the 1950s as garment factories producing for the national market relocated from other regions of the country to use the inexpensive labor of Mexican and Chicano women. Many of these workers did not even reside in the city, but commuted daily from Ciudad Juarez, across the Rio Grande in Mexico (Mitchell 1955, 80–81 in Tower 1991). Since the 1980s, however, the factories started to move operations to Mexico, Central America, and Southeast Asia. Eventually, as already mentioned, El Paso became the city with the largest number of certified layoffs related to NAFTA relocations in the country. Paradoxically, the decline in manufacturing in El Paso has sharply contrasted with the growth of maquiladora plants in Ciudad Juarez, a few miles across from the international divide. This growth in maquiladora plants in Ciudad Juarez represents significant opportunities for entrepreneurs and professionals in El Paso. Aggregate and, at times, urgent demands for supplies, inputs, and services for maquiladora operations offer fertile ground for business start-ups. In addition, technical and administrative upgrades in the industry enhance the opportunities available for engineers and administrators. These opportunities expand the local middle and upper classes, while their polarization with the poor and the working poor also expands.

The rapid growth of the maquiladora industry in Ciudad Juarez brought its own challenges. In the late 1980s, the maquiladora industry was at risk due to serious infrastructural problems. These problems ranged from transportation bottlenecks and power shortages to high turnover rates among workers and the need for sophisticated education in technical and managerial fields. These concerns were reinforced during the NAFTA discussions, due to expectations of an increase in volume operations and greater facility of companies to relocate their production lines to Mexico's interior. The deficits in infrastructure were more complex than basic services and were related to the new requirements of global production. In addition to the technical requirements of transportation, communication, and basic energy supplies, adequate schools to prepare workers, managers, and other professionals to operate at diverse levels of proficiency were required. Some level of comfort is also required to attract other key workers and professional networks to the city and to maintain them. Public and private resources were mobilized to address urgent problems, if only to assure the short-run operational reliability of the region for global production. "La Vista Grande," the public relations activity organized by Ms. Torres in association with the Maquiladora Managers' Wives Association, was part of these efforts. This activity reflected the specific perception of some entrepreneurs, exemplified to some degree by Ms. Torres, of ongoing global changes and the local challenges and opportunities these have created.

GENDER, ENTREPRENEURSHIP, AND CHANGE IN EL PASO

Marcela Torres exemplifies the possibilities of relative success for Latinas in the intensified integration of the local economy into globalized flows. Coming from a working class background, Torres holds a B.A. in communications and is fully bilingual. In the early 1980s, in her mid-twenties, she started a communications firm in equal partnership with another prominent Latina businesswoman, Dr. Candy Mendez. The firm catered to the needs of businesses and government offices generated by the increasingly direct integration of the region into international markets (an increase greatly influenced by the region's accelerated industrialization), and the trade growth leading to NAFTA. The entrepreneurial partnership evolved from their professional contact in 1986, when Mendez ran for a position on the board of trustees of the local community college. Torres, who later invited Mendez to join her in establishing a communications firm, directed her successful campaign.

Their decision to take advantage of the business opportunity was influenced by experiences of gender discrimination in professional promotions, recognition, and retribution. "I saw that women were always teaching men, who then would move on to higher positions while the women remained in the same position," Dr. Mendez said. Ms. Torres added that when she was the public relations manager of a newspaper, her boss refused her requests for salary raises on the grounds that "she was single and did not need the money." She countered that single male employees received more salary raises and their requests were, to her knowledge, never rejected on personal grounds. In her view, the professional contributions of women were rewarded not in terms of their own value, but rather in terms of outdated stereotypes benefiting the employer only. In my first interview with Torres and Mendez in 1992, they expressed the satisfaction of no longer having to confront gender discrimination, at least in their own firm. At the time, Torres was enthusiastic not only for the business opportunities at hand, but also for the potential changes ahead: "the beauty of it is that old rules do not apply. Everything is changing and we must make the most of it." In our last interview in 1997, Torres's enthusiasm persisted, although it was far more contained. She reported that gender discrimination, while not absent, has steadily decreased through the years in the business environment. However, she felt that not enough Latinas were taking the risks required to start their own businesses.

Torres is a conscientious businesswoman with an evident sense of loyalty to El Paso. She is often involved in activities that she feels promote the city's advancement. Though she has been in business for only a decade and is only in her mid-thirties, Torres is a prominent role model and an accessible mentor to other Latinas who are contemplating getting into business or advancing in their professional careers. She received the Small Business Administration regional award in 1992 for her advocacy and support of minority small businesses. Torres's involvements are influenced by a wider vision, a "Vista Grande." Her efforts promote the consolidation of a local entrepreneurial group that could redirect the city's economic growth. In her view, El Paso's current problem is that it has been "sold cheap." She commented that city officials have never had a coherent plan for economic development. Instead, she added, the local economy has been controlled by two powerful business groups who have promoted the city as a low-wage haven for national and international companies. This shortsighted strategy of "bringing jobs to the city" ignored the developmental and environmental implications of the companies it attracted.

Torres's disappointment with the current orientation of the local economy fuels her involvement in the promotion of the maquiladora industry and NAFTA, which she feels could, if properly fomented, bring better economic opportunities to the city in the form of professional and technical positions with higher skills and better wages. This view influences her sympathetic, yet ultimately matter-of-fact, attitude toward displaced workers:

> One worker I met at one of my presentations … said, "you are taking my job away and I have been in [company's name] twenty years." We spoke in Spanish because she didn't speak English. I said, "well, you worked [there] twenty years, how much of an increase in wages have you had over that period?" and she had basically [only] wage increases according to what was being increased by law. [She] never had any retraining or learned to speak English because they keep her there for that purpose. They figured, who else is going to hire her. But how do you tell somebody, "it's wrong what they have done to you and now you have to change." This may be coming across as being cold-hearted, but in fact that's what I have tried to explain to her. She really needs to look at this as a natural change, an evolution that is happening with or without our help. (Interview with Torres, July 1993)

For Torres, the worker is both incorrect in blaming the maquiladora industry for her loss of employment, and misguided in her strategy for solving her problem. In the entrepreneur's view, most of the eliminated garment jobs had been lost to automation or relocations. Most of the relocated jobs, she pointed out, had not gone to Ciudad Juarez, but to Southeast Asia, Mexico's interior, and Central America. Challenging the worker's complaints, Torres argued that El Paso would have benefited had the jobs indeed been moved across the border due to the companies' monetary inflows. Their operational expenses would have stimulated other local economic activities, such as construction and business services ranging from janitorial companies to law and customs offices. These multiplier effects would have also produced direct wages and salaries, most of which are eventually spent in El Paso. Instead, with the relocation of the plants to distant regions, these economic benefits were lost along with the employment gains that they would have generated for El Paso.

Given her interpretation of the processes leading to the displacements, Torres felt that the workers had taken the wrong approach in addressing the problems caused by the relocations: In an increasingly integrated global economy, in which companies could find many other cities around the world with lower production costs, the workers' demands to keep the low-skill, low-wage production operations in El Paso were neither plausible nor sound. Instead, she thought the workers and the government should implement the training required to be competitive in the new global arrangement so as to upgrade the local labor market, its wage levels, and the overall technological standing of the economy. She felt that her business contributed to this upgrading of the local economy through the activities she fostered and the higher-skill jobs she offered to her employees. According this logic, the scarce and rather obsolete developmental contribution of the garment industry was not something to be fostered.

From Torres's perspective, the workers' situation, while acknowledged, is relativized by the "Vista Grande" that she promotes. The workers' displacement appears as part of an inevitable series of transformations that not only extended beyond any individual or specific place, but also involved everyone:

> It's going to happen because we as consumers want to buy products that are competitive to our pocket book. We are going to be looking for ways of finding cheaper places to manufacture and it's our own fault. If we had a way of being able to buy American and pay the prices then we could say, "let's close our borders." But then again, is that smart? I mean, economics tells us it's not smart because that's what Mexico tried to do for so many years. They closed their borders to competition and now you have suppliers that can't even compete in their own country for maquila [contracts]. (Interview with Torres, July 1993)

GENDER, CHANGES, LAYOFFS, AND WHOSE CITY IS THIS ANYWAY?

Carmen Rocha, the director of *La Mujer Obrera* (Working Woman), was not as convinced by the inevitability of this new economic context. In particular, she disagreed with the idea that the garment industry was a thing of the past for El Paso. For Rocha and the other members of this labor advocacy group, the economic shift promoted by Torres and other entrepreneurs failed not only to address the urgent needs of workers, but also to capitalize on their substantial skills. Thus, Rocha felt that, beyond the entrepreneur's expectation that the government take responsibility to retrain workers according to the demands of the emergent economy, much more was needed. She acknowledged the need to reformulate local economic arrangements to respond to the requirements of globalization emphasized by Torres. However, for Rocha, the emphasis should not be solely on retraining the workers, but on enhancing local industrial capabilities to revitalize the garment industry and make it competitive in the reconfigured apparel markets. Rocha's strategy had evolved through years of attending to the immediate needs of workers.

La Mujer Obrera was founded in 1982. In 1991, at the time of our first interview, the group consisted of seven permanent staff members and a handful of occasional volunteers. All but two members of the organization were women. With the exception of two Anglo women, all were Chicanos, two of whom had immigrated to the United States as adults from Mexico and had worked in the garment industry. The organization, which operates mostly with the aid of grants from national and international organizations, designed an alternative strategy to address the disruptive effects of local relocations and the restructuring of the garment industry at the national level. At the time, Rocha was in her mid-thirties. She was college educated and came from a family with a history of involvement in the Chicano movement of the 1960s and 1970s. Her views reflected an ample knowledge of the social impacts of technological and organizational trends reshaping the garment industry as well as of academic and media discussions of these trends.

Along with increased geographic dispersion, the major garment companies reconfigured their corporate structure into subcontracting pyramidlike arrangements. In El Paso, the fragmentation of larger firms created opportunities for former managers and supervisors to open their own subcontracting firms. On the one hand, these firms provided business opportunities as well as much-welcomed employment options in the face of closures. On the other hand, these business opportunities were undermined by impending crises due to the capital and technical challenges they were ill-prepared to meet. In general, subcontractors had a decreased participation in profit margins due to their lower positions on the subcontracting pyramid. In addition, they contended with erratic work orders due to the industry's volatile markets (Tower 1991). In El Paso, the costs and consequences of these problems were often passed on to workers in the form of poor working conditions and below-minimum wages. These irregularities were further accentuated by overnight closures in which the owners left to avoid their financial responsibilities to the city, banks, and workers, including back wages. According to Rocha, some shop owners soon reopened and registered under new names to evade their legal responsibilities. Aware of the weak technical and financial position of small contractors, Mujer Obrera looked for a more cooperative strategy to address the root of the problem: the technical and spatial reconfiguration of the garment industry.

Rocha's direct exposure to the everyday functioning of the industry allowed her to recognize the overwhelming magnitude of the changes and to opt for a collaborative solution. This effort was grounded in a distinction she perceived among entrepreneurs. She differentiated between "*empresarios*" (entrepreneurs) and "*barbajanes*" (scoundrels).

> [Barbajanes have] no business plan or ethics. [They are] those who lack a sense of planning, who set up a business just for the quick profit, with no commitment or sense of building anything, just for the quick kill. They don't care about anything, their business, their workers, the community, nothing. As soon as the business stops yielding a huge and easy profit, they fold down and look for some other chance to make a lot of money with very little work or risk. They just keep going from one business to another, like hustlers or gamblers.

Rocha used the term "empresarios" to refer to those business people who, in contrast with the barbajanes, "have a business plan and care about their business and the community. Those who at least pay minimum wages and care about the basic rights of their workers. These are the business people who are committed to the growth of their businesses, who invest with long term goals in mind. [Those] who understand that they can't just take from the community without putting anything back. Unfortunately, there aren't many of these." Rocha's idea for a collaborative initiative received some support from a handful of garment factory owners, a few public officials, and other concerned business people (*El Paso Times* 1990). However, this cooperative attitude was frequently weakened by practical and ideological factors such as heavy schedules and workloads, as well as a certain amount of mutual distrust. Eventually, they agreed to develop a plan for a concerted effort among the city, the private sector, and workers to capitalize on the long-established assets of El Paso as a garment production center. The plan, which was presented to city officials in October 1990, proposed the development of local infrastructure to match the garment industry's new technologies and processes.

The initial phase of the development strategy was the creation of an industrial incubator, the High Fashion Institute, meant to foster the stability and growth of small firms in the reconfigured industry. The aim was to provide assistance to subcontractors in a very competitive environment. Rocha stated plainly that the goal was to help them become "better bosses" through technical training in bookkeeping, marketing, financial programs, and legal responsibilities toward workers. The success of these small firms, it was hoped, would allow them to specialize in increasingly more sophisticated operations to assure the industry's local stability, as well as higher skill and income levels for workers and the city as a whole.

The High Fashion Institute opened in 1994 with one full-time staff person, two consultants, and a ten-member advisory board, including owners of garment factories and a representative of the largest garment company in the city. The only direct and enduring connection between the institute and La Mujer Obrera was the participation of one of the consultants in both groups. This rather tenuous connection did not seem to concern members of La Mujer Obrera, who stated that they never aimed to incorporate or be incorporated by the institute. The institute surpassed its original goals in each of its first two years of operations, offering assistance to 15 companies each year with approximately 290 workers benefiting from it. However, it failed to promote the diversification of locally manufactured garment products from mostly denim jeans to operations catering to women's fashion markets. Subcontractors were not able to venture into new operations due to financial, technical, or marketing limitations and after four years of operations, the institute ceased its operations.

The hard-won accomplishments of La Mujer Obrera in creating the High Fashion Institute was also dwarfed by wider changes in the local garment industry. At first, garment employment rebounded from its lowest point in October 1990 with 15,800 workers, to its highest peak ever in September 1995 with 21,100 workers. Obviously, this rebound cannot be attributed directly to the institute's accomplishments. Despite its constructive vision, the small scale of the institute's operations and slim resources could not have produced such a major turnaround in such a short amount of time. Rather, employment growth can be attributed to expansions made by larger companies in specialized operations, particularly in cutting. In addition, there was significant growth in industrial laundries doing stone and acid treatments of fabric for prewashed jeans. From this highest figure, however, the jobs in the industry declined again to 18,842 in 1997 (Vargas 1998).

Job fluctuations in the industry overwhelmed La Mujer Obrera's advocacy work. It was forced to dedicate most of its effort to procuring retraining funds granted by the labor clauses of NAFTA for displaced workers. Besieged by an increasing number of displaced and demoralized workers, the group concentrated its efforts on guiding workers through the tortuous process of securing compensations as well as denouncing the deficient training that most workers receive. In 1996, these denunciations led to a six-month extension of benefits for workers who had been affected by initial delays in the training programs and in the government's processing of worker benefits.

Rocha's involvement in La Mujer Obrera has decreased in the last six years. In 1993, she left El Paso to become a board member of a national foundation located in New York City, which promotes community-based projects aiming to advance social justice. The following year, at the onset of the indigenous uprising in Chiapas, Mexico, Rocha became the official spokesperson of the Zapatista Army in the United States. Her departure has not meant a termination of her involvement with La Mujer Obrera, even though it reduced her immediate input and the amount of energy she could directly commit to local events. Her husband, however, has remained directly involved in the organization and played an important role in securing an extension of the relief funds. When I spoke with Rocha's husband right after they had secured the additional funds, he expressed satisfaction for the accomplishment but was very restrained with his optimism for the situation of the workers and the group itself. In recent months, La Mujer Obrera had reduced its personnel to two full-time staff members, further limiting its capacity to address the growing demands of displaced workers.

CONCLUSION: ON THE SUNNY SIDE OF A BLAZING CHANGE

Global changes reshape the lives and opportunities of women in El Paso with a skewed distribution of costs and benefits. Some of these women are ready to operate in an integrated world economy while others are left behind. The dramatic contrast between the large number of displaced workers and the unprecedented public success of a small number of local women exemplifies the kind of perplexing relationship between individual breakthroughs and the collective deterioration of conditions observed by Greer in her assessment of gender inequality in the current moment of increasing globalization. The cases of Torres and Rocha show two energetic and resourceful women actively involved in reshaping conditions in their city. Both women strove to alter institutional arrangements in accordance with new requirements imposed by a globalized economy. At the same time, the different orientations of their involvements and accomplishments reflect the nature of the global economic reconfiguration, in which options and opportunities are unevenly distributed among local actors. For Torres, not only was the garment industry bound to relocate, it was also unable to offer the developmental potential to move El Paso beyond its traditional "low wage" condition. In accordance with her own firm's activities, she advocated for a new economic role for the city based on service and technical activities related to the ongoing regional industrialization process. Her strategy, however, paid little attention to the immediate needs of the displaced workers, which were precisely the focus of Rocha's involvement.

The practical disappearance of the local garment industry in El Paso and the steady growth of the maquiladora industry in Ciudad Juarez give greater credence to Torres's outlook than to Rocha's strategy. Moreover, Torres's business success seems to corroborate the accuracy of her interpretation of developing trends. In 1997, at the time of our last interview, she celebrated the firm's first decade in business. Her success, however, reflects profound challenges that bear witness to the perplexing type of advancements in recent breakthroughs by women alluded to by Greer. After a decade in business, and worn out by demanding interactions with problematic clients or prospective clients, Torres chose to focus on a single product. Ironically, the trying demands placed on her and her firm were generated by the avalanche of consultations that resulted from the business enthusiasm generated by NAFTA. Torres said she was the victim of her own circumstances:

> I go on and I do a presentation to 200 people. Suddenly those 200 people start thinking: "God, the opportunity is so great. I am going to call [Marisa Torres] and I know she can help us." They don't even know what they want to do, but they want to do something. I had a guy call me, "you know, I just sold my sheep ranch and I want to do something in Mexico." He has the money, but I can't think of what I can do to be able to make money as a company and do something for that man. He was going to invest $1 million in Ciudad Juarez. I am not a financial adviser.
>
> You have this type of thing. It has really escalated since this whole media hype on NAFTA. That is our problem right now. ... I get a lot of calls and I am working, but I am not making a lot of money. ... I spend the whole day returning people's calls and what's happening is, since we love to do work and we know that we are good in the information business, we end up taking each one of these calls into consideration and even if you give half an hour of your day to each one, that shifts the whole day for us and one, maybe, out of twenty is a good deal. (Interview with Torres, July 1993)

The considerable amount of time, resources, and calls led Torres to eventually narrow the scope of her business. However, it took time for her to recognize the need for this shift and even longer to implement it, in part because of the alluring potential level of business suggested by a high volume of inquiries and her small firm's initially tenuous position. Some of the inquiries, while vague and outside her field of expertise, promised generous offers of profit sharing or commissions that lured her into preliminary explorations. On one occasion, she estimated that she invested close to $10,000 in phone calls, faxes, and time dedicated by her and her staff. This investment produced no return as the business venture did not materialize.

Torres, like the subcontractors in the garment industry, faced an innovative and volatile market in which ostensible opportunities also created a constant vulnerability to being exploited. Torres estimated that only one of every twenty calls produced any business but there was always the potential that the next call would make up for those that did not yield a profit. Moreover, there was always the hope that such a call would possibly propel the firm along a path of greater success. Because of its recent establishment and its limited capital reserves as a small business, the firm could not afford to blindly shun alternative sources of business. Depending initially on a few temporary contracts, she contended with different types of contingencies. In addition, the apparently favorable market conditions imposed challenging demands on her skills and judgment due to nonmarket contingencies derived from her location, her ethnicity, and her gender. In an interview in 1994, Torres said that she often found herself having to negotiate the price of her directories because of prevailing misconceptions:

> [Some clients] also perceive Mexico as being cheap. This goes back to the way the whole media presented the disparity of wages [between the two countries]. Yeah, you look in the papers: "it is cheap there." No doubt about it. So you have a perception that everything is cheap, including my work: "You charge $100 for that ... directory. I can get that information for free." So pretty soon, I am sitting there and negotiating: "what if I sold it to you for $80?" and I cheapen my own service.

While Torres's firm is registered as a U.S. company, is located in the United States, and most of the companies in its selling directory are U.S.-based, the fact that the maquiladora plants are in Mexico, her firm is on the border, and she is Mexican American were misconstrued by some clients as part of the bargains to be gained from "Mexico" via NAFTA.

Some of these complications suggest the reconstitution of gender subordination noted by Elson and Pearson (1984). As a woman, Torres contended with unprofitable requests and consultations that most likely would not have been made of a man. These unconventional requests were naturalized in the image of the nurturing woman. The requests involved calls from individuals with vague ideas for possible business ventures that they would noncommittally explore with Torres, relying on the perception and unspoken expectation of women as "patient and understanding," that is, "good listeners." Applying the insights offered by Elson and Pearson (1984), the calls can be seen as manifesting business interactions that became "bearers of gender": The new sites and roles of interaction are reinvested by the unequal set of expectations and prerogatives of traditional male–female relations.

The calls also suggest the extent to which Torres herself was trapped in this commonsense state of affairs and point to the insidiousness of the recomposition of inequality, despite breakthroughs. With an implicitly motherly tone, Torres characterized the callers as "dreamers" and "opportunists." "Dreamers" were those without capital, but intensely driven by an innovative business idea to take advantage of NAFTA. "Opportunists" were those with capital, but who would manipulate her into doing work for them under the promise of a business deal that would not come into being: "They end up paying me $20 for something I could have probably sold for $1000," noted Torres. In both instances, these clients also took advantage of her good disposition, which without a doubt led her to offer nurturing advice to the misguided "dreamers" and to trust the manipulative "opportunists."

Her characterization, however, also had a class dimension. Both types of problematic clients took advantage of the firm's weak standing as a new small business, of Torres's enthusiasm for the potential opportunities, and of her friendly disposition. Torres eventually became less patient with the idea that "boys will always be boys"; she has narrowed the scope of her business and is less tempted by vague inquiries with undefined promises. She has managed this by consolidating her firm's operations on the development and sale of its main product, a directory of maquiladoras, the sales of which have consistently followed the expansion of the industry. In ten years, the directory's sales have increased from 500 to 6,000 directories worldwide and it is currently available electronically through the Internet. Counting on gradually expanding revenue from the directory sales, Torres realized that "a lot of calls does not mean a lot of business."

The consolidation of her firm led to her enthusiastic outlook. However, the limited and tenuous resources at her command, considered in light of the potential opportunities, suggest important class considerations easily overlooked if one focuses on the dramatic contrast between Torres and the workers. Considered in terms of the socioeconomic recomposition of the new economy, in this new frontier, entrepreneurship may be acquiring a new meaning, given the new conditions and options for procuring a living. At a time when local mass employment in factories and in government is decreasing in El Paso and in the country as a whole, self-employment becomes less optional or voluntary, as the quantity, quality and stability of salaried jobs decreases. In this transition, Torres's involvement (as an empresaria according to Rocha's typology) in promoting the reconfiguration of the local economy toward a new and improved standing in global markets is meritorious with regard to her attention to social problems and opportunities. However, this attention is constrained to a restricted focus, one in which the pressing conditions of displaced workers is minimized.

Along with this class reconfiguration, important recompositions in the sites—literally and figuratively—of gender inequality underscore the perplexing relationship between individual breakthroughs and collective setbacks. Hopefully, the individual breakthroughs of women like Torres and Rocha will help consolidate and extend the inroads toward gender equality. In the meantime, hopefully, their successes can lead to solutions to the worsening of socioeconomic inequalities overlooked in "La Vista Grande." In such a "big picture," the perplexing coincidence of the worsening of gender inequality with individual women's breakthroughs needs to be recognized in the integrated common grounds of globalization. These globalizing dynamics entail specific interplays of localities and the complex, concrete involvements of women as gendered actors facing greater instability at the individual and institutional levels. This instability calls us to desalambrar our conceptual frameworks in order to step outside ideological or epistemological constraints that hinder our capacities to make sense of the baffling yet dramatic changes in gender, class, and place that globalization generates.

NOTES

Fieldwork for this research was partially funded through a faculty fellowship in the Great Cities Institute of the University of Illinois at Chicago. Janise Hurtig, Manuel Arellano, and Kristine Bartizel provided important comments and editorial support.

1. All names used in this chapter are pseudonyms, with the exception of "Twin Plant Wives Association." The actual name is provided because of its specific gender and ethnic resonances and because its name appears in references provided in the chapter.
2. NAFTA is a trade agreement to eliminate tariffs and other commerce barriers between Mexico, Canada, and the United States. It went into effect in 1994. In 1997, the Texas Employment Commission certified that 6,906 workers in the city lost their jobs because of NAFTA-related relocations. Nearly two-thirds of these laid off workers had been employed in the garment industry (Vargas 1998, 5).
3. These claims draw upon Frederick Engels's influential analysis of the subordination of women in *The Origin of the Family, Private Property, and the State,* originally published in 1884. In Engels's view, the subordination of women under capitalism derived from their economic dependence on men. Thus, for Engels, the gender issue was ultimately a class issue (1972, 173).

REFERENCES

Baake, Ken. 1990. "On 'Maquila' Celebration, Some Are in Mourning." *El Paso Herald Post.* April 20, 1.

City of El Paso, Texas. 1992. Census Profiles From Summary Tape File 3A. Department of Planning, Research, and Development.

Coyle, Laurie, Gail Hershatter, and Emily Honig. 1982. *Workers of Farah: An Incomplete History.* El Paso: Reforma, El Paso Chapter.

Elson, Diane and Ruth Pearson. 1984. "Subordination of Women and the Internationalisation of Factory Production." In *Of Marriage and the Market: Women's Subordination Internationally and its Lessons,* edited by Kate Young, Carol Wolkowitz, and Roslyn McCullagh. Boston: Routledge and Kegan Paul.

El Paso Herald Post. 1990. "Maquila: La Vista Grande." March 19.

El Paso Herald Post. 1990. "Garment-Factory Protestors Released from Jail." June 30, p. B2.

El Paso Herald Post. 1990. "Fed Probe Finds Violations in Local Garment Industry." August 10, A1.

El Paso Times. 1990. "Plan Aims to Keep Garment Industry from Coming Apart at the Seams." May 2, A1.

El Paso Times. 1991. "Bill Makes Non-payment of Wages a Felony." May 28. B6.

Engels, Frederick. 1972 [1884]. *The Origin of the Family, Private Property, and the State.* London: Lawrence and Wishart.

Fernández-Kelly, María Patricia. 1983. *For We Are Sold, I and My People: Women and Industry in Mexico's Frontier.* Albany: State University of New York Press.

Greer, Germaine. 1999. *The Whole Woman.* New York: Knopf (distributed by Random House).

Martinez, Oscar J. 1996. *U.S.–Mexico Borderlands.* Wilmington, DE: Scholarly Resources Inc.

Massey, Doreen. 1994. *Space, Place, and Gender.* Minneapolis: University of Minnesota Press.

Peña, Devon Gerardo. 1997. *The Terror of the Machine: Technology, Work, Gender, and Ecology on the U.S.–Mexico Border.* Austin: University of Texas Press.

Perrucci, Carolyn C., Robert Perrucci, Dena B. Targ, and Harry R. Targ. 1988. *Plant Closings: International Context and Social Costs.* New York: Aldine de Gruyter.

Reich, Robert. 1991. "Who Is Them?" *Harvard Business Review* 67(2): 77–89.

———. 1992. *The Work of Nations.* New York: Vintage Books.

Russell, Joel. 1998. "Sisters, Inc." *Hispanic Business* 20(5): 32–35.

Sanchez, George. 1993. *Becoming Mexican American.* Cambridge: Cambridge University Press.

Stoddard, Ellwyn R. 1987. *Maquila: Assembly Plants in Northern Mexico.* El Paso, TX: Western Press, University of Texas at El Paso.

Tiano, Susan. 1994. *Patriarchy on the Line: Labor, Gender, and Ideology in the Mexican Maquila Industry.* Philadelphia: Temple University Press.

Tower, George. 1991. "El Paso's Colonias." Ph.D. diss. University of Arizona.

Twin Plant News. 1990. "Monthly Magazine on the *Maquiladora* Industry." El Paso, Texas.

Vargas, Lucinda. 1998. "El Paso's Labor-Mismatch Dilemma." *Business Frontiers,* Issue 1. Federal Reserve Bank of Dallas, El Paso Branch.

10. "Making a Scene" ∽

Travestis and the Gendered Politics of Space in Porto Alegre, Brazil

Charles H. Klein

> *At school with my friends, at recess and in the classroom, my desire for men awakened. I did many chores at home, like washing the car and the dishes. At night, I studied the multiplication table. The boys called out to me, but I was afraid. ... My grandmother restricted my activities a lot, so I went back and forth between school and home, without having friendships, because at night I had to study for tests and sleep. I used to go to the circus a lot with my parents on Sundays—it was the only interesting thing that existed for me. I met a gay man who told me about parties, clubs, adult movie theaters, and the street. I never did anything sexual with a woman. They advised me to use hormones, which led me to run away from home. I got to know the night, the street—I stole from cars in order to live. I continue working in the batalha and using hormones. Effect: breasts. Changing the body: As soon as I became a travesti, I received silicone implants. I'm not sure about my future, but I would like it to be the best possible. And to continue to be a travesti until I die!*
>
> —Text by Cristina Loira (blond Cristina), GAPA travesti work groups
> Porto Alegre, June 14, 1994

On the night of July 21, 1994, Cristina Loira, a travesti[1] sex worker, was murdered by a john on the streets of Porto Alegre's main travesti prostitution zone. That same night, Heidi, another travesti who witnessed the killing, reported the crime and the identity of the killer to the police. As is often the case when travestis are killed in Brazil, the police did not conduct a thorough investigation, and no one was arrested or tried for the murder. Given this police inaction, and the fact that Cris was but one of a number of travestis who had been murdered in the past 12 months, a group of travestis based out of the Support Group for AIDS Prevention (GAPA)[2] decided it was time to take the issue of violence against travestis to the people of Porto Alegre. Over the next few weeks, several GAPA staff people, in conjunction with the travesti groups, organized a coalition who staged a protest march against violence of more than 80 people through downtown Porto Alegre on August 23, 1994.

To understand this political mobilization it is necessary to consider the many ways in which AIDS has impacted travestis in Porto Alegre. Many Brazilian travestis are HIV-positive or have died as a result of HIV infection, and travesti sex workers' situations make them especially vulnerable to HIV infection. Since the late 1980s, when the AIDS epidemic began to accelerate, HIV-related risk reduction, discrimination, and services have been of utmost concern to travestis in Porto Alegre. GAPA, Porto Alegre's first and largest AIDS-related organization, began working with travestis soon after its establishment in 1989. Since 1992, this GAPA/travesti relationship has coalesced in GAPA-based groups addressing issues related to AIDS, health, and the *batalha,* that is, the "battle" of working as a prostitute.

My involvement with GAPA dates to 1991, when I first visited Porto Alegre in the process of selecting a site for research on AIDS activism. Upon returning in 1993, then GAPA president Adelmo Turra sat me down. "Charles, last time you were here you were getting to know Porto Alegre and GAPA. Now, you have to make yourself useful—you can't just watch things and take notes for twelve months. You need to become part of a work team and share the skills you have to offer." Turra's suggestion coincided with my vision of "action research": I had come to Brazil not simply to learn about AIDS and sexuality, but to make whatever contribution I might to global AIDS activism.

Turra and Suzanna Lopes, the coordinator of the travesti groups, asked me if I would be interested in cocoordinating the groups with her. I nervously agreed, and much to my joy, soon found out that the travestis at the GAPA groups enjoyed talking about sex and gender as much as I do. No doubt, being a queer North American added to my acceptance by group participants, as we shared both an attraction to men and a fascination with comparing the experiences of *bichas* (fags) in Brazil, the United States, and beyond. But while I consistently was amazed and amused by the travestis' scathing critiques of the Brazilian gender and sexual orders, I was horrified by the violence they reported on an everyday basis. Cris's death was a particular wake up call for me as I, unlike most of the travestis—and many of the GAPA staff and volunteers—had never known anyone who had been murdered. In this situation, it was rather easy for me to *desalambrar* and put aside questions of "objectivity" and "traditional" research methods. So rather than limit myself to understanding the forces behind Cris's murder and the mobilization it generated, I channeled my energies into helping organize the protest march and raise awareness about violence and discrimination against travestis.

The GAPA travesti groups and the mobilization they organized in response to the murder of Cris Loira provides a lens for examining the growing politicization of travestis in Brazil in the 1990s. Examining the gendered spaces in which travestis live, organize politically, and "desalambrar," I highlight how particular spatialized social relations have shaped travesti identities and facilitated the development of a queer performative strategy of "making scenes" that combines microtactics of resistance with mainstream political claims-making.

A BRIEF HISTORY OF GAPA

The history of GAPA/Rio Grande do Sul begins in 1987 in Porto Alegre, a city of about 1.5 million people in the southernmost state of Brazil. At this time, Gerson Barreto Winkler, a data processor and social-work student then in his late twenties, took an HIV antibody exam after his partner began to manifest an HIV-related illness. The test came back positive. Shortly thereafter, Winkler's partner, a physician, entered into a steady physical decline, and the couple was forced to deal with a series of frequently insensitive and misinformed health professionals. These experiences motivated Winkler to work on AIDS-related issues in Porto Alegre. As Winkler explains, "when he died I wanted to do something to change the things I saw at the hospital and to help HIV from spreading throughout the city."

GAPA quickly emerged as the most important space for people with or affected by HIV/AIDS in the Porto Alegre metropolitan area. Many GAPA volunteers were also participants in Porto Alegre's gay/homosexual subculture,[3] which, combined with the group's outreach to transgender and female sex workers, gave GAPA a decidedly nonmainstream reputation as a place for *bichas loucas* (crazy queens), *fechação* (campiness), and social marginality in general. From 1992 to 1993, the campiness and craziness gradually diminished, partially in response to criticisms from a growing cohort of female volunteers who questioned whether the group's ambiance was misogynist and exclusionary. During this same period, the Ford Foundation awarded GAPA its first institutional grant, and the drive toward a more professional work environment was re-enforced with the division of GAPA activities into three divisions—the Social Assistance Center, the Education and Information Center, and the Prostitution Studies Center. GAPA has continued to receive support from international agencies, including the MacArthur Foundation and the World Bank (through the Brazilian federal government's National AIDS Program).

I have called this form of AIDS-related politics practiced by GAPA and other Brazilian AIDS-related organizations "professionalized activism" to highlight the coexistence of a self-described political militancy and a bureaucratized and project-driven organizational structure.[4] The "professionalization" of Brazilian AIDS NGOs differs from most Brazilian private corporations and governmental agencies, which are typically heterosexual male-gendered spaces. In contrast, there have been relatively few heterosexual men among GAPA volunteers and staff, and the women and gay/homosexual men who have led GAPA have worked to create a space in which individuals of normative and nonnormative genders and sexualities can "be themselves" without suffering reprobation and discrimination. Moreover, GAPA encourages women and men to talk openly about sexual activity, which the group presents as a way of being personally and socially responsible. This valuing of sexual communication works against Brazilian gender ideals and erotic culture in which differently sexed partners are not encouraged to talk about sex and in which women (but not men) who demonstrate too much interest in sex risk losing "respectability" (see, for example, Goldstein 1994; Paiva 2000).

As a result of these internal dynamics, GAPA can be seen as a kind of "queer corporation" (Patton 1997) in that it relies on bureaucratic organizational models yet has a political ethos of challenging gender and sexual hierarchies. Nowhere is GAPA's queer corporatist persona more evident than in its travesti groups, where since 1992 a series of GAPA paid staff and volunteers have provided a structured space for travestis to reflect on their genders, sexualities, and experiences. At GAPA, travestis take a break from the *batalha* (sex work) in an environment where many of the intragroup hierarchies (i.e., the "beautiful" over the less "beautiful," the young over the old, the more well-to-do over the less well-to-do) that characterize travestis' lives in other contexts are diminished (see also Benedetti 2000). This downplaying of intragroup differences has encouraged travestis in the GAPA groups to focus upon their commonalities, such as their gender, sexuality, occupation, and concern about AIDS and violence.

Over the years, and with the active encouragement of GAPA staff, this sharing and reflecting has generated a new political identity among group participants, namely, that of the travesti sex professional (*profissional de sexo*), a term affirming that prostitution is a form of work deserving the same labor rights, working conditions, and respect as any other legitimate occupation. This discourse of work and respect has resonated strongly with many travestis in Porto Alegre. In emphasizing the important role that the GAPA groups have played in promoting political awareness and activity among travestis in Porto Alegre, I do not mean to imply that travestis' lives on the street are apolitical. As I discuss below, through "making scenes" (*dando escándalo*) (Kulick 1996, 1998), travestis actively respond to and resist the many injustices they suffer. Yet, it is at spaces such as the GAPA work groups and, more recently, national travesti conferences and travesti rights organizations (Kulick and Klein, forthcoming 2003), rather than the streets, where travestis have begun to articulate collective political strategies that move beyond individualized tactics of resistance.

THE GAPA TRAVESTI WORK GROUPS

GAPA's work with travestis dates to the group's inception in 1989, when volunteers began distributing condoms on the streets of Porto Alegre's main transgendered and female prostitution zones (see Klein 1996). To obtain free condoms and information about AIDS, many travestis began to stop by the new GAPA downtown office, one of the few spaces in Porto Alegre where travestis felt welcome on their own terms. This travesti presence was not favorably received by all who worked in the federal office building in which GAPA had its offices, resulting in a petition calling for eviction. When GAPA relocated to its current offices in early 1991, the association of GAPA and travestis was cemented in regularly scheduled, afternoon "work groups," the activities of which were largely determined by GAPA staff and volunteers. Increasingly in demand, by early 1994 what had been one group was split into three, each with 15 to 25 regular participants, who in exchange for attending a session received between 50 and 75 condoms.

Guiding GAPA's prevention and education activities is the idea that effective HIV risk-reduction programs require exposing and responding to the sociocultural and political economic factors that shape individuals' willingness and ability to incorporate safer sexual practices into their everyday lives. In the case of transgender sex professionals, this project is particularly complicated and involves the "arduous work of deconstructing the stigmas of marginalization and criminality so present in these groups" and "stimulating self-esteem and promoting citizenship and self-determination" (Lopes 1995, 5). One of the primary strategies to achieve individual awareness and group solidarity is to focus on what it means to be a travesti. As can be seen in Cris Loira's autobiography above, in addition to bodily transformation in spite of prejudice, travestis cultivate "access to a network of information and the appropriation of knowledge, finger nails, hair, eye brows, clothes, make-up, hormones, silicon." Indeed, this is a particular language community: "the entrance ticket to this universe is completed through learning a distinct language, the 'bate-bate'——a code language, of African origin, used by the group. To be travesti is much more than simply a man dressed as a woman" (Lopes 1995, 5).

Since "to be a travesti is much more," before turning to consider how travestis at the GAPA groups perceive and experience sex, gender, violence, and work, it is useful to situate travestis among other cross-dressing and transgendered persons. Brazilian travestis generally distinguish at least three categories of male to female "transvestites" by the degree to which the individual is positioned along "male"/"female" gender lines. On one end is the drag (drag queen) who dresses up in women's clothing and goes to bars and clubs. The drag (most, but not all, drags are gay/homosexual-identified men) does not consider himself to be a "woman" or "female" gendered, but cross-dresses from time to time for fun. A second category is the *transformistas*, who can be further separated into two groupings. One is the *carnaval* tradition of men dressing as women and taking to the streets (see also Parker 1991, 144–148). In this context, men of all sexual practices and identities *desfile* (parade about) in drag during *carnaval* without suffering social reprobation. A more specific use of the word "transformista" refers to men who perform as female impersonators in (homosexual) bars and clubs. Nearly always subculturalized and self-identified gays/homosexuals, these men are theatrical professionals who dress up as woman, lip sync songs, and tell off color jokes.

A travesti is neither a drag nor a transformista, although some travestis do work in clubs or consider performing a possible career option outside of prostitution. What most distinguishes travestis from the other two categories of *travestismo* is the fact that travestis typically maintain a "female" gender for twenty-four hours a day (with clothes and make-up as fits the hour or occasion) *and* physically transform their bodies to achieve their idealized visions of the "feminine." Moreover, nearly all travestis have been or are sex workers, making it difficult to separate travesti and prostitute/sex worker subjectivities.[5] Almost all the travestis I know have expressed limits as to how far they are willing to change their body, with most stating that they would not have a sex-change operation, even if they had the necessary financial resources to do so (most do not). As a result, in English I refer to travestis as transgendered persons, rather than "transvestites" or "transsexuals"; and although one could find travestis at each of these extremes, most fall somewhere in between.[6] For example, Roberta Close, Brazil's most famous travesti, had a sex-change operation in Europe. Close's unsuccessful attempt in the mid-1990s to change her official sex from male to female on her Brazilian documents was much discussed by travestis and the mainstream media. Conversely, Rogéria, another well-known Brazilian travesti, is a vocal opponent of sex-change operations.

In describing travestis as "transgendered," I do not mean to suggest that travesti subjectivities in Brazil are the same as those of male to female transgendered individuals in North America and Western Europe. Unlike many male to female transgendered people in the United States, who usually do not describe themselves as having a "gay" identity or "homosexual" sexual practices, most Brazilian travestis consider themselves to be *bichas* (fags), albeit of a special type.[7] Some travestis go so far as to argue that they are the "real" fags, with gay/homosexual men who present more "male" genders constituting closet cases who deny their true "female" gender (see Kulick 1998). Nonetheless, like most male to female North American transgendered individuals, and unlike many Brazilian men who adopt "gay" identities, travestis do not usually have sexual or romantic relationships with gay/homosexual identified men, instead preferring gender-dichotomized relationships with "straight" men. These examples highlight the dangers of uncritically viewing Brazilian genders and sexualities through U.S. and Western European lenses, even though Brazilians of non-normative genders and sexualities use terminologies derived from the United States and Western Europe (e.g., "gay," "homosexual," "transsexual") and often explicitly situate themselves within global gender/sexual collectivities.

Judith Butler's (1990, 138) ideas of gender as both performed and material provide a useful starting point for understanding these complexities of travesti genders and sexualities. In the case of travestis, the process of transforming oneself from a boy/man into a travesti who uses hormones and receives silicon implants is, as Butler argues, a personal and cultural history in which imitation and illusion are central elements. Most travestis do not consider themselves to be "women," and they recognize that many women do not act or resemble the at times mythic versions of "women" toward which many travestis aspire. The maintenance of corporal transformations is also difficult, requiring continuous depilation and more often than not the taking of female hormones. As a result, travesti bodies are explicitly fluid. At the same time, when necessary or useful, travestis shift into "male" gendered practices, such as dropping the register of their voice, wielding a knife, or trying to pass as a "regular" man.

This idea of gender as a process of continuous reconstruction was repeatedly raised by travestis at the GAPA groups, who generally described their own transformation as having involved two distinct, though interrelated, processes.[8] The first phase centers on the realization that a boy/young man might be a travesti, the second on the subsequent social processes of constructing a travesti body and assuming a public travesti identity. Complicating the construction of travesti genders and identities is the complicated yet undeniable connection between travestis and male homosexuals. On the one hand, most travestis at the work groups described male homosexuals as having, like themselves, more female psychological characteristics than "real men." According to this logic, man: woman :: man: travesti :: man: bicha.[9] On the other hand, these same individuals noted that many gay Brazilian men are noticeably—and increasingly—different from travestis and "traditional" bichas in terms of their gender presentation (e.g. they are more "masculine" and often sport gym toned bodies) and sexual partner selection (e.g., they are attracted to other gays and prefer non-gender dichotomized relationships). But in the end, nearly all travesti group participants agreed that uncertainties about the etiology of homosexuality and travestis notwithstanding, while one might be born a homosexual, one must *become* a travesti.

And how does one become a travesti? Catarina, in her early twenties, describes the process as follows:

> It all began when I was seven years old. I still was a boy and played with other boys—soccer and other games.
>
> When I was ten, everything began to change. I had female cousins and they invited me to play and I went to play with them and they let me. I played with dolls. I loved to make clothes for them.
>
> When I was twelve I met a travesti on the way to school. The travesti was very beautiful. But we never spoke. And then one day she came and talked to me. I sat down on a bench in the park, and she came up and asked me if I were a bicha. I didn't know what to say. She told me not to be ashamed of saying yes, because she knew that I was a bicha. She began to tell me her life story—both her happy and sad times. And she told me that if I wanted, I could be like her. Beautiful. I said I wanted this. She told me that to be like her I would have to take an injection. But I was afraid of receiving an injection. I didn't see her after this.
>
> I keep studying and always went back to the same place to see if I could meet with her again. Three years went by, and I saw her and she called me over and asked if I had lost my fear of injection. I said that I had and began to take the *chamado hormônio* (the dreaded hormones).
>
> I even lost my fear and shame. The years went on and a friend asked me why I didn't work in the *batalha*. I said that I was afraid. She said that she would always stay with me until I learned enough to do it alone.

Becoming a travesti involves a passage from one's childhood home (the place of gender-coded play) to the frightening yet fascinating street.

Here, older and wiser bichas and travestis introduce potential travestis to gay/homosexual local knowledge and the technologies of being female, including dress, make-up, hormones, and silicon. During this acculturation process, the teenager learns of the possibility of making "easy" money as a prostitute, and is frequently encouraged by his mentor(s) and new friends to take this route. And the more feminine and pretty s/he is, the more money s/he will likely make in the batalha.

This connection between becoming a travesti and becoming involved in sex work shapes the everyday lives, worldviews, aspirations, and identities of nearly all travestis. For if, as Catarina's friend suggests, sex work can provide material benefits to young travestis, who often lose the support of their birth families when they begin to transform their genders and bodies, this "battle" is no easy game. In Porto Alegre, like most Brazilian cities, travesti prostitution zones are located in out-of-the-way places (e.g., dark parks, streets with little traffic) characterized by substantial petty theft and high numbers of physical assaults. Clients and police officers beat up on travestis, and travestis frequently rob clients when they do not receive what they consider to be sufficient payment for their services. Some travestis also incorporate theft into their sexual-economic exchanges on a regular basis, even if the client has paid the agreed-upon price. In this manner, circles of violence are created in which clients think that travestis are violent criminals, and travestis fear that clients and the police will harass, beat, or kill them. As one work-group participant stated a few weeks prior to the killing of Cris Loira, "I leave home (to work on the streets) and don't know if I will return alive."

It is important to stress the gendered nature of violence against travestis, since it is almost always heterosexually identified men who seriously harm and kill travestis, rather than the other way around. Though travestis suffer the brunt of the physical violence in the prostitution zones in Porto Alegre, travestis are not passive victims. Rather, travestis respond to these situations by "making scenes" (*dando escándalo*), which takes various forms, from robbing clients to fighting in bars over boyfriends to using scathing, off-color language in the face of social discrimination. In most scenes, travestis turn the tables on their supposedly masculine and heterosexual clients, employing the same pejorative language that clients use to condemn and justify violence against travestis. As such, travesti scandals are not so much about contesting the Brazilian social order or promoting a more positive image of travestis, but rather about spreading the shame experienced by travestis (Kulick and Klein, forthcoming 2003). Through making scenes, travestis distinguish themselves from bichas who do not rock the boat *and* provide a potential foundation for constructing a shame-conscious and shame-creative (Sedgwick 1993, 13–14) "queer" alternative to assimi-lationist gender and sexual politics (see Klein 1999; Kulick and Klein, forthcoming 2003).

If these connections between sex (work), the street, violence, and mak-ing scenes are central to the formation and maintenance of travesti social identities, they also structure the ambiguous ways in which mainstream Brazilian society views travestis. For many Brazilians, travestis, especially extremely beautiful ones, epitomize the "feminine," as can be seen by the centrality of travestis within carnival and the fact that Roberta Close was considered to be Brazil's "most beautiful women" in the mid-1980s (see Kulick and Klein, forthcoming 2003). Nor is the sexual desirability of trav-estis limited to a few glamorous celebrities, as demonstrated by the tens of thousands of travesti sex workers in Brazil. Yet, if the femininity and exotic sexuality of travestis are admired and often desired, travestis and the sexual pleasures they offer are seen as dangerous, which, for some, may be part of their erotic thrill.[10] Intensifying this fear in which Brazilians hold travestis is the association of travestis with violence and AIDS, two themes that trav-estis consistently invoke and re-enforce when they make scenes. But per-haps the greatest threat posed by travestis lies in their capacity to break the veil of the "four walls of the bedroom" (where, according to a common Brazilian expression, "anything goes") and reveals that supposedly mascu-line men in fact like to *dar o cu* ("take it up the butt").

The position of travestis within Brazilian society is thus variable and context-dependent. In spaces where travestis' disruptive capacities can be contained or domesticated, as in carnival and drag shows, travestis are usu-ally admired—if not taken seriously—for their exotic sexuality and outra-geousness. More complicated is the position of travestis on "the street," whose seemingly unlimited possibilities afford both great pleasure and pain for travestis. On the positive side, many travestis enjoy sex and do not see themselves as exploited simply as a result of being sex workers (e.g., many travestis will not charge attractive clients [see Kulick 1998]). Nonetheless, for all its erotic possibilities, the street is an extremely dangerous place for travestis, and outside of the travesti prostitution zones—and some, though by no means all, homosexual/gay subculturized spaces—the street does not welcome travestis. For example, many travestis report being denied entrance into commercial establishments (e.g., restaurants, bars, sex motels): "If you are at a bar or restaurant, the police drag you out, hitting you. And you can't wait at a bus stop [made up and dressed in women's clothing]—you have to bring a sack and change" (Heidi). This social discrimination against trav-estis extends to sexual and romantic relationships as well: "I dream of hav-ing a home that is nice, and this should be easier to do than it is. My husband [for four years] can't ask to rent one and all that, because if he did, they'd find out about me, and he might lose his job. So we have to keep the two worlds separate" (Alexandra).

Even walking on the streets is often no simple matter: "It's not only for prostituting that we suffer—we are harassed even when walking on the streets with our mothers and family." (Angela). Living with such discrimination and violence takes its toll on many travestis, as can be seen in the following poem, which was written during an exercise at the travesti work groups a few months before the killing of Cris Loira:

> Fear of the people around me.
> Fear that the condom will break.
> Fear of death.
> Fear of old age.
> Fear of solitude.
> A fantasy of one day finding a client who gives us all we need.
> The illusion of riches.
> The illusion of being more than others.
> Fear of the prejudice of our own clients.
> Fear of the darkness of the streets.
> Fear of my own friends around me.
> Fear of myself.

This poem and the above comments suggest that for all its openness, Brazilian erotic culture is not without homophobia, though with Brazilian specificities.[11] Such homophobia—and the multiple forms of discrimination and violence it entails—structures the everyday lives of travestis, who as a result of their gender presentation and outrageousness, suffer disproportionately in comparison to others with less readily visible forms of gender and sexual non-normativity. But homophobia and violence can also motivate collective political action among travestis, such as the protest march against violence in response to the killing of Cris Loira. In making such "political scenes," travestis effectively turn homophobia on its head and destabilize, at least temporarily, the gender and sexual underpinnings of the Brazilian sociopolitical order.

MAKING A POLITICAL SCENE—MOBILIZING AROUND VIOLENCE

One week before the killing of Cris Loira, the GAPA groups turned their attention to the question of violence on the streets after Cris and Donna, one of the group's newer participants, recounted how they had been assaulted by several police officers a few days earlier. Even though all present agreed that violence was rising in the prostitution zones, substantial differences existed as to what should be done in response to these realities. Heidi worried that reporting these incidents might cause the police to retaliate with even more brutality. Donna, in contrast, audaciously put forth the idea of video recording police abuse in action and sending the tape to television stations. And in a strange moment of foreshadowing, Cris questioned whether any kind of reporting was worthwhile when most people wanted travestis off the street anyway.

A few days later, my cocoordinator, Lopes, received a telephone call from several group participants, who told her a client had murdered Cris the previous night. According to Heidi, who was at the scene of the crime, Cris and a client were arguing in the latter's car about the price of a completed trick. As often occurs, the client did not want to pay, or at least not the price Cris was seeking. He yelled for Cris to leave the car. She refused. He pulled out a gun and shot into the floor, and then point-blank at Cris. Heidi screamed "don't kill her," causing the client to shoot at Heidi as well, saying "get away, or I'll kill you too." He drove away, and a few blocks later threw Cris's body out of the car. Several other travestis found the now-deceased Cris and called the police. According to them, when two police officers arrived, a resident offered to put a sheet over Cris's body. One of the officers plugged his nose and commented "AIDS is in the air"; the other said "there's no need to waste a good sheet—does anyone have any newspaper?" Vanessa, another travesti who works in this area, responded by putting her coat over Cris. Somewhat later in the evening, Heidi reported Cris's killing to the police, the only official report of the crime.

At the first group meeting following the murder, Heidi and the other witnesses recounted the events. Sensing that the opportunity was ripe to mount a collective political response to violence against travestis in Porto Alegre, Lopes suggested that GAPA draft a letter expressing rage at the murder and the lack of an adequate police investigation in response to it. Shortly thereafter, GAPA's secretary general reported to the group that GAPA had filed an official complaint with the state government's Commission on Citizenship and Human Rights, who agreed to look into the matter.

At the next meeting, Lopes reported that the Brazilian Lawyer's Association, the Commission on Citizenship and Human Rights, and various elected officials (nearly all from the Worker's Party, who have run the Porto Alegre municipal government since 1989) had agreed to sign the letter of protest. Several travestis suggested that while a letter was a good start, it was not enough, and argued that GAPA and the travestis should organize a *passeata* (protest march). This idea was favorably received, and group participants came up with three major themes for the protest march: (1) work security on the streets, (2) justice, including the legal punishment of the killer, and (3) violence in everyday life. A general assembly open to all travestis and interested parties was scheduled for the following week to plan for the protest march, at which time the final details were nailed down.

The protest march was held on August 23, 1994, with over 80 people marching from GAPA's offices through downtown Porto Alegre to the *Esquina Democrática* (Democratic Corner). The *Zero Hora,* Porto Alegre's principal newspaper, printed this summary of the protest march and demonstration:

> Travestis hold march to denounce violence—homosexuals count eight deaths in the last ten months. The regular passer-byers at the Democratic Square in Porto Alegre, presently bombarded by political pamphlets, were jolted yesterday afternoon. Instead of campaign materials, they received information sheets denouncing violence against travestis. Organized by GAPA, the demonstrators simulated a funeral in order to mark the fact that eight travestis have been murdered in the capital since November 1993. The last to die, by four shots to the face, was *Cris Loira,* in July. [italics in original]

All those involved in the protest march considered it a great success, both in terms of directing media attention to violence against travestis and of demonstrating that broad based coalitions could be mobilized in support of travesti-related issues. That there may be a space for travestis within mainstream progressive Brazilian politics is further demonstrated by the slogans that travestis carried at the protest march, which included *"bicha não é bicho"* (fags aren't beasts), *"travesti é cidadão"* (travestis are citizens); *"dignidade por todos"* (dignity for everyone), *"prostituição também é trabalho"* (prostitution is also work). These slogans, and their emphasis on work, citizenship, human dignity, and social justice, fit solidly with the political discourse of the Worker's Party and suggest that the "sex professional" identity encouraged by GAPA has taken hold among many travestis in Porto Alegre. Through voicing these claims at Democratic Corner, travestis and their allies inserted travesti-related issues into local and national political discussions on citizenship, social justice, and violence in Brazil (e.g., Fernandes and Carneiro 1995; Velho and Alvito 1996). This expanded travesti participation in mainstream Porto Alegre politics continued in the months following the protest march, with Heidi testifying before the Commission on Citizenship and Human Rights and a GAPA travesti group participant being appointed an alternate representative to the municipal Commission on Human Rights, Violence, and Discrimination.

Yet, despite these seemingly strong connections between the protest march and mainstream Brazilian politics, the newspaper report above does not situate the protest march within either broader political discussion on citizenship, social justice, and violence or the upcoming elections and instead highlights its shocking and disruptive dimensions. That the protest march could "jolt the surprised on-lookers"—and, it would seem, the *Zero Hora* journalist—is largely the result of travestis occupying a place from which they are nearly always absent. Most travesti sex work occurs at night after regular commercial activities have stopped, and it is primarily in this context that travestis present themselves fully *montada* (made-up and dressed provocatively in female attire). In contrast, during the day travestis spend most of their time at home, sleeping and preparing for the next night of the batalha. Should they go outside during the day, travestis usually dress in a casual manner, which, while feminine, does not draw particular attention so as to diminish the likelihood of their being harassed. By taking to the streets and occupying one of the city's main political public spaces (Democratic Corner) in drag, the travestis broke down the social and spatial confinement of travestis within late night prostitution zones. In this way, the protest march "queered" a normally heterosexually organized and sexually understated social space *and* enabled travestis to participate in one of the most traditional forms of Brazilian public political claims-making.

Seen in this light, the protest march against violence is a "big scene" that destabilizes spatial and social categories according to a similar logic as the microlevel scandals discussed above. Much as individual travestis verbally disparage their clients and shame them into paying more by exposing their "femininity" and "homosexuality," so in the protest march the performance of fully montada travestis demanding their citizenship in broad daylight in Democratic Corner shames mainstream Brazilian society into considering travesti grievances. In moving from a micro to a collective political scene, travestis link the shame of sexual hypocrisy (i.e., many of the very men who were shocked by the protest march desire and/or pay for sex with travestis) to the shame of societal complacency in the face of continuous violence and discrimination against travestis.

Might this political tactic of combining "making scenes" with traditional claims-making offer a strategy through which travestis can demand full citizenship within the Brazilian sociopolitical order? If so, how might achieving citizenship affect the everyday lives of travestis, given that "making scenes" (e.g., using verbal abuse and threats of physical violence against clients to further travesti economic objectives) seems incompatible with most conceptions of citizenship? Might becoming citizens ironically result in travestis losing their ability to make scenes, thereby depriving them of a tried and true means of defending themselves against social and economic injustice? Or might it be possible for travestis to redirect the focus of their scenes from sexual economic exchanges with individual clients to more collective efforts, such as the protest march against violence, and use shame to directly contest gender, sexual, and economic inequalities?

How might travestis work toward achieving citizenship without losing either their identity or ability to use shame to defend their interests? As the mobilization in response to the murder of Cris Loira demonstrates, AIDS activists have been important allies of travestis, and given the ongoing impact of HIV on travestis, these partnerships are likely to continue in the future. Yet, if queer corporatist AIDS groups such as GAPA have provided an important foundation for travesti organizing, including a growing national travesti "movement," the resulting travesti political movement has simultaneously become extremely dependent on AIDS-related funding (see Kulick and Klein, forthcoming 2003). For example, in 1999, a group of travestis in Porto Alegre founded *Igualdade* (Equality), the city's first travesti rights organization. Igualdade "works in defense of the rights and citizenship of travestis and transsexuals ... and has developed various activities, such as biweekly discussions about human rights, citizenship, self-esteem, and health prevention (STDs/AIDS)" (Igualdade brochure 2000; note the similarity to the themes and organizational structure of the GAPA work groups). To date, all funding for the group has come from the National AIDS Program at the Ministry of Health. Similarly, the annual "National Meetings of Travestis and Open-minded People Who Work With AIDS," which is the centerpiece of national travesti political organizing, is totally depended on AIDS-related funding from the National Health Ministry and international philanthropic agencies (see Kulick and Klein, forthcoming 2003). These realities suggest that the continued intermeshing of travesti political organizing within AIDS organizations and the AIDS industry, though understandable given the scarcity of financial resources available for travesti-related issues, may limit the scope of travesti politics and impede the development of autonomous, self-sufficient travesti organizations and political movements.

How might travesti organizing ease its dependency on the AIDS industry? One possible strategy would be for travestis to form partnerships with gay, feminist, and other organizations, as occurred in the mobilization around the killing of Cris Loira, since such groups arguably share with travestis a mission of challenging gender and sexual inequalities. Yet, most gay and feminist groups have not incorporated travestis into their political discourses, agendas, and actions other than the occasional exceptional event (e.g., the protest march against violence). Rather, most Brazilian gay groups have tended to emphasize the (homo)sexuality of travestis and to subsume them into a monolithic category of (male) "gays/homosexuals" without paying much attention to the gender issues raised by living as a travesti, while many Brazilian feminist/women's groups have operated according to a logic in which gender (at least in practice if not in concept) is equal to (biological) women, thereby leaving travestis entirely out of the picture.

In highlighting these obstacles to integrating travestis and travesti-related issues into other social movements, I do not mean to imply that travesti political action should be or is somehow limited to participation in formal political organizations and structures. For example, in recent years, a national gay culture has been emerging in Brazil (see Klein 1999; Parker 1998). It may prove possible for travestis to use these cultural forums to explore the relationship between travestis and (male) homosexuality and to create new images and collectivities that do not subordinate travestis. Exactly how travestis might fit into these gay spaces remains uncertain, particularly since many travestis, at least at the level of ideals, operate according to a binary gender logic of male/female and male/travesti oppositions that many Brazilian gay activists seek to destabilize. From this perspective, travestis, both through their social practices and perhaps their very existence, may be seen as reenforcing, rather than subverting, the dominant gender and sexual orders.

It is likewise not clear whether the existence of a consumer-oriented Brazilian "pink economy" bodes well for travestis and other Brazilians of non-normative genders and sexualities. On the one hand, the pink economy may be more about making money than working toward social justice; on the other hand, given the increasing dominance of a market-oriented, neoliberal order in Brazil, the growing economic clout of (at least some) "gay" Brazilians may translate into some kind of political power. Participation in Brazil's emerging gay worlds also requires a certain level of disposable income, and the more entrenched the pink economy and its "gay consumer ethos" become, the more likely that those who are poorer— and travestis are typically low income—will have a limited capacity to participate in these spaces. And along with the pink economy often come ideas of respectability that may exclude travestis and other "queers" whose gender and social behavior rock the boat.

This being said, it is important to remember that "gay" communities and politics in Brazil, as in other places, are by no means homogenous, and a sizable number of nontravesti "queer" Brazilians are questioning the gender, sexual, and economic underpinnings of Brazil's emerging gay communities. Queers themselves also come in many forms, from the activists at Nuances,

Porto Alegre's principal "gay" rights organization, who explicitly reject assimilationist gay politics, to national queer cultural critics, such as Jackson A. and Erika Palamino, who have pushed the boundaries of gender and sexuality in their writings for the *Folha de São Paulo*, Brazil's leading newspaper (see Klein 1999). Travestis, particularly those who continue to make scenes both large and small, have an important role to play in these efforts to queer Brazil, and their outrageousness—and contradictions— might help prevent Brazilian sexual and gender politics from falling into assimilationist complacency. Seen in this light, travestis and their shame-conscious and shame-creative strategy (Sedgwick 1993) of making scenes not only promote an "intimate citizenship" that affirms "our most intimate desires, pleasures and ways of being in the world" (Plummer 1995) but also provide a promising model for political action in a contemporary world characterized by globalization, shifting identities/subjectivities, and the shameful effects of neoliberal economic policies.

NOTES

I thank the travestis at GAPA and Igualdade for graciously sharing their thoughts and experiences, GAPA for the opportunity to cocoordinate the travesti work groups, and Don Kulick, Barbara Hobson, and the entire Recognition Struggles project at Stockholm University's Gender Institute. All translations are mine.

1. Travesti is a Portuguese term used to describe certain male to female transgendered individuals. A more detailed discussion of travestis and how they fit into Brazilian gender and sexual categories is provided below.
2. The first *Grupo de Apoio e Prevenção à AIDS* (GAPA) was founded in São Paulo in 1985. Since that time, GAPAs have opened in many Brazilian cities. Although sharing the same name, the GAPAs are independent organizations. In this chapter, I use the acronym GAPA to refer to GAPA/Rio Grande do Sul.
3. The words *gay* and *homosexual* in Portuguese resonate differently than their English counterparts. *Homosexual* is generally less medical in feel than "homosexual" in English. *Gay* is more explicitly political than "gay" in English, and in the mid-1990s had some of the feel of "queer" as used by activists in contemporary North America. The use of colloquial terms such as *bicha* and *veado* (faggot, sissy) and *sapatão* ("big shoe," dyke) is also widespread throughout all Brazilian social classes. "Gay male" activists in Brazil generally refer to themselves and their community/movement as *gay* and/or *homosexual;* Brazilian women sometimes use the term *lésbica*, which has an explicitly political connotation. Many Brazilian men have sex with other men without considering themselves to be *bichas, gay, homosexual,* or *bisexual*. In this chapter I use the composite "gay/homosexual" to note the instability and changing nature of these words and social categories.

4. Klein 1996; see also Galvão 2000 and Paulson, chapter 6, this volume.
5. Klein 1996, 1998; Kulick 1998; Kulick and Klein, forthcoming 2003; Parker 1998.
6. Many scholars and activists use the term "transgender" to incorporate different kinds of people who cross gender lines (Bolin 1993; Califia 1997; Hausman 1995).
7. Klein 1996, 1999; Kulick 1998; Kulick and Klein, forthcoming 2003.
8. See Benedetti 2001 on gender construction among travestis in Porto Alegre.
9. See Kulick 1998 for similar ideas among travestis in Salvador, Bahia.
10. See Parker 1991 on the role of rule breaking in Brazilian erotic culture.
11. On homophobia in Brazil, see Braiterman 1998; Larvie 1997, and Parker 1998.

REFERENCES

Benedetti, Marcos. R. 2001. "*Toda feita:* O corpo e o gênero das travestis." Master's diss. Universidade Federal do Rio Grande do Sul, Instituto de Filosofia e Ciências Humanas, Program de Pós-Graduação em Antropologia Social. Porto Alegre, Brazil.

Bolin, Ann. 1993. "Transcending and Trangendering: Male-to-Female Transsexuals, Dichotomy and Diversity." In *Third Sex, Third Gender: Beyond Sexual Dimorphism in Culture and History,* edited by Gilbert Herdt. New York: Zone Books, 447–486.

Braiterman, Jared. 1998. "Sexual Science: Whose Cultural Difference?" *Sexualities* 1(3): 313–325.

Butler, Judith. 1990. *Gender Trouble: Feminism and the Subversion of Identity.* New York: Routledge.

Califia, Pat. 1997. *Sex Change: The Politics of Transgenderism.* San Francisco: Cleis Press.

Fernandes, Rubem C. and Carneiro, Leandro P. 1995. *Criminalidade, drogas e perdas econômicos no Rio de Janeiro.* Rio de Janeiro: ISER.

Galvão, Jane. 2000. *AIDS on Brasil: A agenda da construção de uma epidemia.* Rio de Janeiro: ABIA and Editora 34.

Goldstein, Donna. 1994. "AIDS and Women in Brazil: An Emerging Problem." *Social Science and Medicine* 39(7): 919–929.

Hausman, Bernice. 1995. *Transsexualism, Technology, and the Idea of Gender.* Durham: Duke University Press.

Klein, Charles H. 1996. "AIDS, Activism, and the Social Imagination in Brazil." Ph.D. diss. Department of Anthropology, University of Michigan.

——. 1998. "From One 'Battle' to Another: The Making of a Political Movement in a Brazilian City." *Sexualities* 1(3): 329–343.

——. 1999. " 'The Ghetto is Over, Darling': Emerging Gay Communities and Gender and Sexual Politics in Contemporary Brazil." *Culture, Health and Sexuality* 1 (3): 239–260.

Kulick, Don. 1996. "Causing a Commotion: Scandal as Resistance among Brazilian Travesti Prostitutes." *Anthropology Today* 12(6): 3–7.

———. 1998. *Practically Women: The Lives, Loves, and Work of Brazilian Travesti Prostitutes.* Chicago: University of Chicago Press.

Kulick, Don and Charles H. Klein. Forthcoming 2003. "Scandalous Acts: The Politics of Shame Among Brazilian Travesti Prostitutes." In *Recognition Struggle: Social Movements, Contested Identities, and Power,* edited by Barbara Hobson. Cambridge: Cambridge University Press.

Larvie, Patrick. 1997. "Homophobia and the Ethnoscape of Sex Work in Rio de Janeiro." In *Sexual Cultures and Migration in the Era of AIDS: Anthropological and Demographic Perspectives,* edited by Gilbert Herdt. New York: Clarendon Press.

Lopes, Suzana H. Soares de. 1995. "Tornar-se travesti, ser cidadão." *HIVeraz* 3: 5.

Paiva, Vera. 2000. "Gendered Scripts and the Sexual Scene: Promoting Sexual Subjects Among Brazilian Teenagers." In *Framing the Sexual Subject: The Politics of Gender, Sexuality and Power,* edited by Richard G. Parker, Regina M. Barbosa, and Peter Aggleton. Berkeley: University of California Press.

Parker, Richard G. 1991. *Bodies, Pleasures and Passions: Sexual Culture in Contemporary Brazil.* Boston: Beacon.

———. 1998. *Beneath the Equator: Cultures of Desire, Male Homosexuality, and Emerging Gay Communities in Brazil.* New York: Routledge.

Patton, Cindy. 1997. "Foreword." In Brown, Michael P. *Replacing Citizenship: AIDS Activism & Radical Democracy.* New York: Guilford Press.

Plummer, Kenneth. 1995. *Telling Sexual Stories: Power, Change and Social Worlds.* London and New York: Routledge Press.

Sedgwick, Eve K. 1993. "Queer Performativity: Henry James's 'Art of the Novel.'" *GLQ* 1(1): 1–16.

Velho, Gilberto and Marcos Alvito, eds. 1996. Cidadania e violência. Rio de Janeiro: Editora UFRJ.

11. By Night, a Street Rite ∽

"Public" Women of the Night on the Streets of Mexico City

Marta Lamas
(Translated by Lessie Jo Frazier)

The concept of desalambrar put forward by this volume helps us, first, to think about the ways in which fields of proprietorial power are constituted—that is, how fences are built, patrolled, and maintained in relation to challenges from those defined as outside; and second, to consider how those defined as outside permeate the perimeters in actions ranging from going over or under the fences to acts of appropriation such as squatting and poaching to the out-and-out over-running and dismantling of those fences. In framing how terrains of privilege and power are demarcated and challenged, I am particularly interested in the fact that subordinate people inhabit spaces defined in part by their own domination in often ambiguous and contradictory ways.

In the contemporary urban Mexican context, night and street together form a pivotal place, understood in this case as an intersection of spatiality with temporality, and constituted culturally as a masculine domain. Within that domain, sex work, whether performed by women or men, is constituted as a feminine position, especially as those who purchase these services are almost exclusively "men." However, does this feminized position mean that sex workers always occupy a place of submission, of passivity? This is among the questions that led me to engage in a collaborative research and political organizing project with sex workers.[1] In this chapter I draw on my fieldwork "in the street" to examine how these sex workers use notions of freedom and control as well as ritual to resignify the place they occupy. I will argue that the active (dis)positions of many sex workers and the ways in which they manage their laboring lives are far from social positions of victimhood; on the contrary, they are positionings that subvert dominant definitions of gender.

Marc Augé (1993) develops the concept of the "anthropological place" as a conceptualization of space that expresses cultural references. In this sense, the place where urban Mexican sex workers offer their services on the street—referred to as the *punto*—is the epitome of an anthropological place. It is a small space, symbolically marked, with observable, definite limits. My goal in this essay is to show how some sex workers subvert a dominant understanding of the night and the street as realms of masculine power as well as dominant definitions of their own gendered positions through two aspects of street sex work in "el punto": First, through their ideas of freedom and control over work; and second, through the ritual performed by sex workers to appropriate the street as a place of work and, in the process, to protect themselves. I conclude by returning to the initial question I posed about agency to suggest ways of thinking about the sex workers as occupying a place not only of stigma but also of transgression.

THE STREET AND THE NIGHT

That the street at night is a place of masculine power is a belief still so naturalized in urban Mexican culture that women who (inappropriately) occupy that place become responsible for any consequences that result from their action. This is exemplified by the words of Judge Gustavo Aquiles Gasca in a criminal trial that occurred in Mexico City. He denied legal recourse to Claudia Rodríguez Ferrando, a mother of five who wounded an attacker who subsequently died due to lack of medical attention. Blaming Ms. Rodríguez for the man's death, the judge argued that "that was not a time [*esas no eran horas*] for a decent woman to be out on the street" (cited in Domingo, June 2, 1996). Such conceptions expose the vulnerability of the position of women who work the streets with respect to the patriarchal law. While the logic of this lesson was clear to the women with whom I worked, it deserves more careful attention here.

According to dominant ideologies, the gendered cultural logic of the street is such that those women who remain in the domestic sphere are chaste, honest, and decent, while, conversely, those women who walk about on the street are wayward, indecent, or whores. For this reason, I find eloquent the notion "public woman" to refer to a prostitute as "the woman available to all." The concept—profoundly rooted in our society—of the public woman has been part of a distinct moral code for women and men that has been so naturalized that its logic seldom requires elaboration. Thus the division between "decent" women (*mujeres decentes*) and whores (*putas*) carries a spatial dimension by reproducing in social space gendered structures of domination. The valorization established by the sexual stigma of whores serves to keep in line all women. Bourdieu (1988) states that the prescription of the double sexual moral code is interwoven with an asymmetrical status ascribed in an economy of symbolic exchange. Thus invested with this symbolic function, women are forced to continuously work to preserve their symbolic value, adapting themselves to masculine ideals of virtuous females while augmenting their value and attractiveness with cosmetic applications and other bodily endeavors.

In urban Mexico, the structural pinnacle of female value is motherhood, the feminine ideal expressed by the sequence femininity–maternity–love–service–abnegation–sacrifice. Note that, symbolically, mothers are not (openly) sexualized: The Virgin Mary is the ultimate representative of this idea—to be a mother without having had sexual relations. Desexualization comes to be the normative aspiration of the feminine ideal. This paradoxical message is converted into a sanction against women who exercise their sexuality with pleasure (or for the satisfaction of other needs) and from there it is a short distance to the next step: to being stigmatized as whore. In point of fact, the insult "whore" is directed at women who depart from this narrow moral norm and have sexual relations without reproduction and outside of marriage, whether or not their sexual activity is commercialized (see Montoya, chapter 3, this volume, on "bad women" in Nicaragua). Thus, sexual repression has pushed enjoyment—and satisfaction in the broadest sense—to the dark side of femininity. This argument can also be extended to understand that any work not directed toward social reproduction in the domestic sphere similarly can place women in the category of "whore," whether it be sex work or some other wage labor outside of the home, though clearly this implication has become less overt in recent times.

Having looked briefly at the spatial component, in this case the street, let us now consider the night. Place also has a temporal dimension, and for our purposes we shall consider the night as constituting a particular component of the relevant place in our case study, el punto. While I noted earlier that the night is considered a domain of masculine power, especially in terms of the prerogative of male mobility, this observation needs to be qualified in relation to the cultural symbols of femininity associated with the night in Mexico (e.g., gendered cultural sentiments about the moon or the qualities of darkness). There seems to be a cultural contradiction expressed in the symbolization of the night as a feminine space and the prohibition against certain kinds of women (or female gendered bodies) circulating in the night. In Mexican culture, the night has been the realm of bad femininity in that it can encourage in women characteristics dangerous to men: Witches circulate at night and at night women can emerge with unbridled, voracious sexuality. For this reason, although the night is symbolically feminine in many ways, it is a place controlled by men who retain for themselves the prerogative of mobility or the right to penetrate the darkness, and by men who seem threatened by the potential (sexual?) power of women. Beyond the dominant understandings, for women in Mexico, as in many other places, the night carries a very real sense of risk, vulnerability, and exclusion.

Given the flood of Mexican women entering the formal labor market since the 1970s, the terms of valuation of women who move about on the street have undergone modification. To the extent that it has become socially acceptable for a "decent" woman to work, thus requiring her circulation on the street, I would argue that the daytime street is currently being cast as more benign or open to women. In other words, we see an even greater bifurcation of spatial distinctions between the daytime and nighttime street as two distinct places. Thus, the ultimate labor taboo for women is nighttime work, only permitted by Mexican law since 1974 (the year Mexico modified a number of discriminatory legal codes in the face of the International Year of Women). The continued social prohibition against nighttime labor acts as another way of sanctioning a gendered social division: to protect women from the risks of the night was, supposedly, to protect them from the danger of sexual violence, or even more so, their own uncontrollable sexuality. This dominant rendering of the nighttime street as a particularly dangerous place for women to inhabit, though no longer codified in law, is clearly maintained today in juridical practice, as in the legal case with which I began this section.

The paradox is that in a changing context in which women occupy more public spaces and widen their range of action and mobility, an archaic restriction persists that, in effect, preserves the night as a sphere of masculine sexual violence. The protected privileges of male sexuality reflect the social hierarchy of gender: the position of power that "men," as a group, have over "women" (and over subordinate, "less male," men, such as homosexuals). Generally, men do not live in risk of rape by women (although certainly by other men), and their position as subjects who enjoy the liberty to use their own body does not stigmatize them because of their sexual activity. The relations of power that objectify women, however, lend themselves to the punitive commodification of the female body. For this reason, the commercialization of sexual activity—prostitution—can elucidate structures of power that give form to the social valuation of sexuality. The conflicts that surge forth in relation to the circulation of female bodies in spheres that are supposedly masculine—the street and the night—are quite connected with the use of the female body in supposedly feminine spheres, epitomized by the home. In both domains, the home and the street, extreme forms of abuse and sexual violence emerge. Thus, we should not think that the transgression lies solely in the commercialization of sex; rather, it is the exercise of feminine sexuality that generates discomfort and conflict for men. Let me be clear: the "female" body as an object for masculine sexuality is no danger; however, women who exercise agency in their bodily practices and movements through space/time are profoundly dangerous, and attractive. For this reason, I focus here on some experiences of women sex workers who occupy this most dangerous of gendered places, the nighttime street.

CHOOSING THE STREET

To understand why sex workers would choose the street as a place of work, we should take a moment to quickly review the organization of sex work in Mexico, tracing prostitution as a practice constitutive of the historical development of boundaries of public and private. Mexico has tolerated prostitution as a so-called necessary evil at least since the sixteenth century with the express authorization of the Spanish Crown to construct a bordello in 1524.[2] Tolerance lasted until the nineteenth century, when sex work was regulated from a sanitary perspective, imitating the French regulation. This prompted measures that included the formal registration of workers (including noting whether their form of work was "independent" or in a brothel). This was followed by the government designation of a hospital in Mexico City that exclusively attended prostitutes. The regulatory system opened the door to coercion, abuse, and corruption among sanitary authorities and the police. The revolutionary years of the early twentieth century saw a dramatic increase in rates of venereal disease and thus government efforts at sanitary control. In 1926, a sanitary bureaucrat claimed that there were some twenty thousand prostitutes, of whom only two thousand were healthy, and that, ostensibly as a consequence, more than half of all Mexicans suffered from syphilis. Lazaro Cardenas's government took new steps to control the sex industry, inspired by the International Abolitionist Federation based in Geneva, legally abolishing prostitution in Mexico City, and with it, formal sanitary regulation. Today, although prostitution as such is not prohibited, the legal situation varies from state to state. In Mexico City, the penal code only criminalizes pimping and procuring, while the Regulations for Police and Government (*Reglamento de Policía y Buen Gobierno*) consign sex work to a moral infringement of good manners or decency. It is this regulation that has been used to control prostitution. With regulations in hand, round-ups are staged supposedly to monitor the health of the sex workers. This use of the regulations in contrast to the actual law has meant the ongoing exploitation and mistreatment of sex workers on the part of judicial authorities and the police, and has lent itself to all sorts of abuses of power without representing an effective form of sanitary control. Clearly, prostitution in Mexico has long been treated as a public matter under government jurisdiction in the name of moral sensibility and public health, thus crossing and helping to constitute gendered borders of public and private. Sex workers' historically shifting and currently ambivalent relations to the law renders them always vulnerable and on the margins of formal public processes of legal recourse. Consequently, the spatiality of sex work takes similarly ambiguous and interstitial forms.

In Mexico City today, there are five basic, clearly distinguished spatial locations for sex workers:

1. Common brothels (*prostíbulos populares*) located in some markets
2. Public spaces (*vía pública*) on the street (with access to short-term hotels)
3. Bars, night clubs, cabarets (with access to hotels)
4. Massage parlors (*estéticas*) with on-site services
5. Call girls (apartments and hotels).

The socioeconomic zone where the exchange happens determines the type of prostitution that flourishes there. In high-class residential neighborhoods, workers won't be on the streets, nor will bars or brothels disguised as estéticas. Instead, there are elegant apartments where the "call girls" (using the English term) attend their clients while a waiter or bartender serves as guard and cashier. In more populated sectors, there are frequently brothels and short-term hotels (*hoteles de paso*) where streetwalkers take their clients. The middle class enjoys a wider range of options: Hotels where streetwalkers go, massage parlors in commercial and tourist zones, and some call girls. The geography of commercial sex work is structured according to an active and competitive market, where the rates are determined not only in relation to the type of service offered, but also according to the beauty, age, social class, and ethnicity of the woman. The combination of all of these factors lends the dynamics of supply and demand a wide range of possibilities. The sites of sex work locate and cross boundaries of public and private, or covert and audacious locations. One of the most contradictory and transgressive of these places is the nighttime street.

In Mexico City it is not possible for a woman to stand around freely to look for clients on the street without being noticed by the authorities or the competition. Work on the street is organized by ferocious territorial control, as much on the part of the authorities as by those who control the women, previously known as *padrotes* (pimps) or *madrotas* (madams), today called *representantes* (agents). Only in certain sectors of the city do the authorities "tolerate" street prostitution. Although territorial rights to work the puntos are assigned by seniority among those who work the street, they also represent the power of those who control the women. Control is an indispensable element for negotiation with the authorities, who must respond to the neighbors' complaints. If a woman wants to work on the street, she has to enter a group that already has a "representative" or agent who negotiates with local government authorities and police.

In the sector of Mexico City I researched, the workers give the agent 50 percent of their earnings and the agent takes care of everything: He or she goes to the town hall for them, handles the bribes, pays the bills, and so on.[3] There is a sort of rotation system among the women because they tire of the agents as much as of the clients. More lasting are the relations among colleagues. When the women meet a work colleague they like in the punto, they frequently establish intense personal relations: They become godmothers of their colleagues' children, get engaged to or marry their brothers, and so on.

The *chicas* (girls)[4] establish contact on the street with the clients, who parade before them in cars, and less frequently in this district, on foot. The negotiations are verbal, and if they agree, they arrange to meet in a nearby hotel. Among the duties of the agent is to arrange for drivers who take the chicas to the hotels; the chicas have to pay the driver the *dejada* (fare). This system was put into place after several clients perpetrated violent incidents of kidnapping and abuse. Since then very few chicas dare to get into a car with a client. The *vestidas* ("dressed" or transvestites) will do so as, in the face of the risk of HIV infection, they have reinstated the custom of giving *mamadas* (blow jobs) in the clients' cars. (See Klein, chapter 10, this volume, for a case of Brazilian travesti sex workers organizing around issues of street violence and HIV prevention.) This new form of work means, moreover, that the client cannot check the genitals of the vestida, which means that he can sustain his fantasy that he's dealing with a "woman." The AIDS epidemic has introduced heightened rivalry between chicas and vestidas, now that clients no longer run the risk of an "exchange of vital fluids" and prefer oral sex.

In general, among women who work on the street, those who work at puntos are of a higher socioeconomic and educational level than those who work in the clandestine brothels[5] and of a level similar to those who work in bars and salons. Why do some decide to work in the street while others prefer an enclosed space such as a salon or apartment? An apparent advantage of working in apartments or massage parlors seems to be greater protection by the police and less risk of harm by the clients: Many more street workers are murdered than chicas who work "inside." However, according to women with whom I spoke, the advantage of street work is, in their words, "freedom." This expresses both the freedom to "choose" clients (street negotiations permit workers to reject whomever they dislike or "gives a bad vibe") and a greater freedom to work on the days they choose. For this reason, there are already puntos where the workers have to pay a fixed fee, independently of whether it covers the amount of services they provide or whether the worker shows up for work. The agents who use this stricter system are those who are on the better streets. They argue that the fixed quota is needed to impose discipline because the chicas on these streets work whenever they want to and neglect business or are too picky about clients.

Street work also implies a greater danger: a continuous confrontation with the authorities and dealing with groups of drunks, including the risk of being run over as the drivers try to drive up over the curb, showing off their machismo or their drunkenness. For this reason, the chicas who elect to work in the street manifest traits of independence and bravery.

Claudia Colimoro, a sex worker, political leader, and agent at the *punto del Oro,* talks about the street as a space of greater labor freedom, but freedom through control over her own sexual labor and the clients' use of her body.[6] Claudia, who worked in apartments and salons, had a late discovery of the street and considers it much better than enclosed spaces, "because [you spend a lot] on cosmetics, on clothes, on taxis to go to an apartment, to go well put-together; then you arrive and you have to drink, you have to stay a minimum of two to three hours until the bottle is finished; then [the client] doesn't come, and you complain to the one in charge or the apartment landlady, and there are times when they don't pay you if they don't come and they make fun of you." Claudia compares this to the street where " ... their 20 minutes are quick and they don't treat you like in the salon, like in the apartments, as in all the places where one works, that they grab hold of you like a float there, they move you, they raise you, they lower you, and then, let's see, a little of 'the French' (oral sex) and they go back to making love to you and, no, no, no a little more of 'the French,' and that's how the salons and apartments are." Claudia further indicates that they can earn more money on the streets, because they charge for each separate thing: "Yes, when you undress completely [you also charge], you assess everything because everything has a separate fee; so you do very well here; the truth is, I regret having worked in the apartments and salons as well, because I thought that in the salons you would have been given certain considerations after all the years you had worked, but no. So I should have known how to work on the street, it would have been better to work on the street all the time, because I would have earned more ... and moreover, here on the street, you know that you'll get picked up by the patrol, but you get out, and in the salon, no." Claudia refers to the situation in which a patrol raids an enclosed place with an order to close shop and all of the workers are taken to the station. Although this occurs with much less frequency than the daily harassment of the authorities on the street, when it happens, there's no simple way of "fixing" one's situation.

Other than the patrols, the major disadvantages of the street are standing on your feet and being in the cold. Claudia comments, "Yes, [with the patrols] at first I was quite nervous, but after a while I became much more brave. That's the worst and the cold, no?, and standing on your feet. For me, my calves began to ache here, I began to have a lot of pain ... my legs, here it's the legs because one isn't accustomed to spend so much time on one's feet; so now you have to shift from one foot to the other because of the chill, and I catch colds/ ... ".

Even after Claudia's initial enthusiasm for the street had worn off, she decided that the street still offered greater freedom and control over her own work, "There are many things, here on the street you learn a lot. I believe that you learn more here and it's much nicer here. You risk a lot of dangers, but, I believe, you also learn a lot. I assure you that had I known about the street, I never would have gotten involved in the salon, nor in the apartment, nor in a bar, nothing. ... It's just that before the street was controlled by pimps, then it was the pimps who would take care of you. You had to be under one in order to be able to work on the street, but no, it was impossible, if we had some here, like it used to be, we wouldn't be in business."

Moreover, there is the complex question of status: Those who work on the street, whether in puntos, clandestine brothels, or bars and hotels, are seen to be lower in the hierarchy: "Everywhere it is seen to be of another category than the salon, the apartments, or rather, it goes by a scale, no?, apartment, salon, and street, but the truth is that we are more foolish, those of us who have worked in apartments or salons rather than on the street, because on the street you don't put up with groping, nor do you put up with a lot of attitude, you don't put up with anything." A little later, Claudia speaks of what seems to her to be the greatest difficulty: "But here on the street, I think that it is very difficult to get clients. In the salon, yes, I had clients who were 18 years old, ever since I began, and they sought me out, more than anything for 'the French' because I did a lot of that."

For the purposes of this essay, I want to underscore some women's sense of themselves as having chosen their place of work according to their own desires, which they describe in terms of freedom and control. In the context of the commercialization of sexuality based on the objectification of the female body as commodity, these workers' sense of themselves as actors who claim prerogatives of mobility and choice constitutes a particularly transgressive stance. The women of el punto instantiate this transgressive and paradoxical stance in a particular collective ritual.

THE RITUAL OF *DEL ORO*

During my fieldwork, I saw how sex workers sprinkled sugar and holy water on a symbolic square in *el punto del Oro*. When I asked Claudia what this meant, she explained that, before beginning work, they make a cross, "On all sides there is a cross made of sugar and holy water is added, or a cross is made of chiles, or one urinates sketching a cross, if you can urinate forming an imaginary cross, all the better."[7] She continued, "Here, imagine!, and with so much traffic!; it is possible that they will take us away for going around making greater moral affronts, because of course we make affronts; for this reason, it is better to bring holy water."

It is a quotidian ritual that represents and accomplishes several goals of which one is to attract clients: " ... daily, when we arrive, we put holy water and sugar and we pray to Saint Judas Tadeo or to Saint Martin Caballero to bring us clients. ... " And this practice is generalized in all settings, "all prostitution has it [this 'custom'] and does it, in apartments with chiles, with sugar, in salons with alcohol, circles of alcohol or crosses of alcohol; yes, in salons, in apartments, in bars. When in a bar there isn't anything [when there are no clients], one exits the bar and the one who is wearing a dress pees forming a cross so that the clients will begin to arrive; everywhere."

When I ask why they make a cross, Claudia diverges to expand on their religious practices, "Prostitution has many beliefs, and amulets, haven't you seen the amulet that they carry? It is an amulet that is sold in [the market of] Sonora, that is cured by the seven forces; the one that I carry is of the Sacred Death (*Santísima Muerte*). ..." The medal to which Claudia refers is a coin with the typical image of death, holding a sickle, carved into the coin; Claudia explains, "Well, I use it because I think that she is the one that takes care of me, death, that is the only thing that we carry with us with any certainty when we are born; [we know] that we are born but the only sure thing is that we are going to die. When? or where?: who knows. And so she protects me so that I will have a good death and so she protects me from all of the people that I have around me, because, well, you know that here you never know which way it will come from, no?, if you are on the street, in an apartment, or in the salon. ..."

The risk of death is a reality in sex work. Claudia remembers how she began to use her medal: "My Santísima Muerte, look, when I had just begun to enter prostitution everyone told me, listen, pray to the Santísima Muerte, but the truth is that I took her off, then they wanted to give me one but I didn't want one. Then, in Orizaba, a girl gave me one. Look, in Orizaba I must have been some, let's see, nineteen years old, I went to work all of the southeast from Orizaba, Córdoba, all, all, all, all, to Isla de Mujeres, I went only to the [red-light] districts; then, there in the district they put their altar to the Santísima Muerte and they pray. There, they told me that they never lack for money with the Santísima Muerte and that, well, she takes care of you, so I caught on and began to carry her, but nothing more than to have her image. Some time passed and afterwards, here, in Mexico, in the jail of Santa Marta Acatitla, they made me one in a coin. Once I discovered that they crafted in coin, from then on I have carried only coin pendants. Furthermore, I have one in my house. ... Yes, pay attention and you'll find that custom among almost none other [than us], or among artists, yes, in artistic circles or in prostitution, because the rest of the people don't believe in death. ..."

When the authorities "accepted" el punto del Oro, Claudia had to obtain holy water to placate the demands of her chicas: "They told me, 'bring sugar and holy water. ...' [W]e went to bring it from Chalma, yes, you see that my daughter was due to give birth, we went to ask that she come through the birth okay; from there, we brought a container for two liters of holy water. The priest there blessed it for us, because here in the churches they are hardly willing to give you holy water, because they know precisely what you are asking it for; ... they are really stingy, they won't give it to you. ... "

The sex workers talk about the ritual of making the cross, sprinkling the place with holy water, or with other liquids or elements, as "following a custom," without making explicit the resonance with a religious (specifically Catholic) identity, such that it seems that they use this mechanism to appropriate a place for themselves. Perhaps the symbols of an accepted identity (the cross, the holy water) codify a rationale of protection. Perhaps the women appropriate for themselves the honorability of the paradigm of a devoutly Catholic Mexican woman, or by the use of a common code of a more general national-religious identity generate a sense of identification that does not separate them so much from the rest of the population and that gives them a certain sense of security.

Belonging to the group "prostitutes" or "sex workers" implies being a part of a collective that cannot openly be revindicated away from its own space. Beyond the particular frontiers of el punto or of the group of professional colleagues, identity should be disguised, negated. Even so, in symbolically marking the ground as their own particular place, some women are generating a situation of identification among themselves, enacting thus a practice common among other collectivities. Hence, another possible angle of interpretation would be that perhaps the ritual serves to elaborate the stigma of their identity as workers. As such, the importance of the rite comes in predominantly subjective terms, as the cultural oppression of the stigma persists in social relations.

Gerd Baumann (1992) suggests that in addition to being an expression of the particularity of a given group of practitioners, rituals are also about "competing constituencies." Furthermore, rituals not only celebrate the perpetuation of social values and knowledge but also express hopes for cultural change. Baumann's central thesis is that rituals, rather than being limited to those who participate in them, are also directed toward "others" and that they serve to negotiate distinct relations. Baumann maintains that the anthropological distinction between "us" and "them" could be a resource of the ritual itself, where ambiguities can be played with and elaborated, and where constituencies can align and realign themselves in a negotiation of who will be "us" and who will be "them." However, in the case of the sex workers of el punto, the distinction between them and other women is a painful and strikingly vivid reality, and so the ritual seems to be directed at subsuming themselves to a more ample "us," specifically, "we Mexicans (women) who pray and use holy water" (since religiosity is a gendered practice in Mexico as in other countries historically influenced by Catholicism).[8] In sum, I propose that through this ritual, a group of women street workers strives to destigmatize themselves.

FINAL THOUGHTS

There are many possible interpretations of this ritual; however, I will limit myself to suggesting a few connections between the cultural symbolism of the street and the night as a productive way to get at a question relevant to women and to society in general: the limiting frontiers—disparate and oppressive—of gender.

David Parkin (1992) affirms that rituals tend to have a directional character—that they are like paths across time and space. He defines ritual as a "formulaic spatiality" carried out by groups of people who are aware of its imperative and compulsive nature and may or may not fill this spatiality with spoken words. From this perspective, what is the transfer that is working in el punto? Could it be that of the transgression of the anonymous woman from apparently being "decent" to being a "whore," and/or that of the invasion of a seemingly masculine space? Men seem to feel threatened when women transgress into spaces allocated to men by gendered power dynamics, the nighttime street being a particularly critical case. Yet men cannot do without "women of the night" since the "night is theirs" *because* of their relationships with these women. In spite of these contradictory dominant cultural schema, the street is a space occupied by both men and women.

The stigmatization incurred by sex workers works as a form of control over women in general in that the fear of being considered a whore exercises a symbolic violence against all women (Lamas 1999). For (supposedly decent) women, this fear is an obstacle to various forms of political and labor participation. The fear of stigmatization is also one of the greatest obstacles to the political organization of sex workers (see Lamas 1993, 1996). Indeed, as society characterizes people's identities, in part, by their sexual activities, the cultural symbolics of gender either imbue with honor or denigrate certain expressions of sexuality depending on whether they are practiced by a woman or a man (or other category) and with whom they are practiced. Thus, sexuality is converted into a means to submit, classify, and humiliate women. The social lesson that sex work offers is a lesson about sexism: Women are marked in a differential manner, negatively, in comparison with the men who purchase their services.

Although it is difficult for a social actor to accept taking on a devalued identity—that of the whore—perhaps there is something positive and gratifying in occupying the street, in circulating by night. Even though I have not elaborated upon this point in this essay, one of the things that made the greatest impression on me in my fieldwork with these women was the way in which many sex workers expressed an enjoyment (*jouissance*) generated by the immense pleasure of transgression, by their interpersonal power over the client, and by the dizzying thrill incited by the risk of living on the edge of marginality. Could it be, then, that the ritual is not only about exonerating stigma, but is also exculpating guilt for this enjoyment? The subversive potential of the ritual might be located in its celebration of initiative by drawing on the contradictions within the most dominant of patriarchal symbologies, the Catholic Church, and appropriating them in an audacious juxtaposition of sacrality and profanity.

If we remember that psychoanalysis establishes a difference between enjoyment and orgasm, we can understand how it might be plausible that many women find personal emotional (psychological) enjoyment from this work, although—according to my interviews at least—they do not find sexual pleasure. This insight opens a whole realm of research and reflection toward which my research with the sex workers of el punto del Oro has just begun to hint. Just as studies have been done of men who seek sexual relations with "whores," because they divide women by a sexist classification, in the same manner there should be a study of the ways in which some women find certain enjoyment if they are placed, objectively or symbolically, in the position of "whore."

This issue of jouissance offers fertile terrain for research, although it requires a psychoanalytical approach. But for now, let us at least conclude that the ritual might express the psychical ambiguities of enjoyment. We need not, and indeed, as anthropologists we cannot, leave our analysis of the ritual vulnerable to psychologistic notions of fundamental intentionality, conscious or otherwise. Parkin (1992) proposes that it is precisely because ritual is fundamentally a physically symbolic action, and the words—as part of the action—optionally replicable, that ritual has a pronounced potential for the "performative imagination," which cannot be reduced to the particular verbal affirmations of a given enactment. This margin of flexibility would allow workers to elaborate particular and varying meanings through ritual.

The women workers of el punto del Oro describe their work there as a choice (and a creation) of a place. In their telling, this place, among a number of possible locales for work, satisfies their needs for freedom and control. Yet, it is a relative liberty in terms of mobility, control over one's own labor, and the dignity of work; and while the risks bring a kind of tough glamour, they clearly entail very real and very probable harm. In such a contradictory context, the ritual of el punto marks this place as one that the women sex workers, even though captives in it, actively shape and negotiate. It is a collective ritual pushing against the stigmatizing and isolated labor structures of the sex industry. And it is a rite that instantiates their need for protection and collective support in a dangerous place. In that sense, it is a ritual that asserts agency.

In el punto del Oro, paradoxical and distinct experiences of femininity conjoin and cross. Perhaps a key aspect of the identificatory process is a certain recomposition of identities and ascriptions, which leads to new models of consciousness and sensibility. From there, the political/labor revindication of many of these transgressors of the limits of "decency" not only crosses culturally valued margins, but also defines the contours of a new feminine sexuality: more egalitarian, although still threatened by men, symbolically speaking. If a key point of the ritual is destigmatization, this becomes even more interesting since there seems to be a tendency in the wider culture of urban Mexico toward the erosion of gendered sexual stigma. So even though stigma works as control, little by little, sexual prohibitions for women are eroding while taking shape generationally is a new phenomenon: steadfast disinterest in maintaining a "decent reputation." Nevertheless, faced with a well-entrenched double moral code, we can be confident that, for the foreseeable future, there will always be a limit to transgress, a margin to press against, a frontier to cross, and a fence to tear down.

The ritual thus becomes a contestation of dominant gender relations in its simultaneous defiant challenge and implicit acceptance of the sex workers' position as defined by those power relations. The ritual is created within the space opened up by the contradictory constructions of femininity in dominant Mexican society, and it has the potential to call such constructions into question to the extent that women sex workers must live out and creatively negotiate those contradictions.

Meanwhile, these workers' activities continue to be a mystery and attraction for the general public, who is scandalized and, consequently, protests, making more difficult a juridical reordering that can facilitate new forms of work organization. For new forms of sex work organization to be effective, there needs to be a public debate, supported by research, that demystifies the inhabitants of the night and shows them as the women of flesh and blood that they are. It is not enough for these women to engage in an implicit cultural criticism through collective rituals of creating and inhabiting places of social connection. To transform implicit criticism and the creation of collective identities into political practices of real liberation, political organizing, always a risky endeavor, must be attained. Perhaps, in this way, a more democratic public debate can foster an open commitment to their current demands and struggles.

NOTES

1. In 1989, I committed myself, as a feminist, to a political organizing project with a group of street sex workers in Mexico City. As a result of this project, in 1990 I was invited to form part of an international research project with the AIDS and Reproductive Health Network. The Mexican component of the project integrated qualitative and quantitative methods including a survey of 914 workers who answered a questionnaire with 120 variables, in-depth interviews, meetings with 8 focus groups, in addition to ethnographic fieldwork for which I was directly responsible. (See Uribe et al. 1990.) The practicalities of doing an

ethnography of this type of place would not have been possible without the assistance of Claudia Colimoro and various other women workers, especially those of the *punto* of the *Calle del Oro.*

2. My historical summary combines a number of sources, the most important of which are the works of Carmen Nava 1990, Salvador Novo 1979, Sergio González Rodríguez 1989, and Bliss 1996.

3. I conducted the initial research in 1990. Still, ten years later, we find basically the same model of control, although in certain small ways the situation for workers has improved now that they can count on defensive spaces, such as the Mexico City Human Rights Commission.

4. This euphemism is used by the authorities and agents to refer to the workers, and the workers themselves use it when speaking of their *compañeras.*

5. The clandestine brothels draw on women with scarce resources, often migrants from the countryside, the great majority of whom are illiterate or have had only minimal elementary schooling. These women walk the streets and take clients inside the clandestine brothel—a large space with cots divided by curtains, often located in a market. The majority of women who work in brothels do not see themselves as "prostitutes," rarely "dress up" to attract clients, and, in general, are timid and shamed. This contrasts with the practices of other, usually younger and more attractive, women who work on the street—in puntos, bars, and salons—and have access to hotel rooms. Among women who work on the streets, these women are regarded as higher in status than brothel workers. From the perspective of sex workers employed in massage parlors and apartments, however, all street workers are lower than themselves in both class and status.

6. Interview, August 1990.

7. In many places, urine is used in rituals to attract men. This also appears in common sayings. In Nicaragua, for example, when a man is infatuated and acting besotted, people say that he acts as if "*le orinaron la cabeza*" (the woman he's infatuated with urinated on his head) (Rosario Montoya, personal communication).

8. Regardless of the level at which anthropological research is applied, it always has as its subject of interpretation an *other.* The sex workers are the first to recognize this division. They were my *other,* and with them, I discovered how *I* was necessarily left to one side by this division.

REFERENCES

Augé, Marc. 1993. *Los "no lugares": Una antropología de la sobremodernidad.* Barcelona: Gedisa.

Baumman, Gerd. 1992. "Ritual Implicates 'Others': Rereading Durkheim in a Plural Society." In *Understanding Rituals,* edited by Daniel de Coppet. London: Routledge.

Bliss, Katherine Elaine. 1996. "Prostitution, Revolution and Social Reform in Mexico City, 1918–1940." Ph.D. diss. University of Chicago.

Bourdieu, Pierre. 1988. "Social Space and Symbolic Power." *Sociological Theory* 7(1).

Domingo, Alberto. June 2, 1996. "Editorial," *El Nacional.* Mexico D.F., Mexico.

González Rodríguez, Sergio. 1989. *Los bajos fondos.* Mexico, D.F.: Editorial Cal y Arena.

Lamas, Marta. 1993. "El fulgor de la noche." *debate feminista* 8 (September 1993).

———. 1996. "Trabajadoras sexuales: Del estigma a la conciencia política." *Estudios Sociológicos* 14(40).

———. 1999. "Violencia simbólica, mujeres y prostitución." In *Antropología política,* edited by Héctor Tejera. Mexico: INAH.

Nava, Carmen, ed. 1990. "Informe de la búsqueda de referencias sobre prostitución en el Archivo General de la Nación." Unpublished Manuscript.

Novo, Salvador. 1979. *Las locas, el sexo y los burdeles.* Mexico: Editorial Diana.

Parkin, David. 1992. "Ritual as Spatial Direction and Bodily Division." In *Understanding Rituals,* edited by Daniel De Coppet. London: Routledge.

Uribe, P., M. Hernández, B. De Zalduondo, M. Lamas, G. Hernández, F. Chávez Peón, and J. Sepúlveda. 1990. "HIV Spreading and Prevention Strategies among Female Prostitutes." Abstract, Seventh International Conference on AIDS. Florence, Italy: W:C3135.

4. Critical Commentaries

12. Against *Marianismo* ∽

Marysa Navarro

Marianismo was a concept first used by the political scientist Evelyn P. Stevens in an essay entitled *"Marianismo:* The Other Face of *Machismo"* (1973a). She defined marianismo as "the cult of female spiritual superiority which teaches that women are semi-divine, morally superior to and spiritually stronger than men." She asserted that marianismo like machismo existed throughout the continent, although marianismo may have been unknown and/or misunderstood when Stevens set out to reveal its importance to Latin Americans and North Americans alike. But it soon found a receptive audience among the latter and beyond and continues to be referenced by scholars writing on Latin American women and even Latina issues.[1]

While marianismo has been criticized, many scholars nevertheless seem to believe that it describes a situation that exists, albeit some may find that Stevens exaggerated its significance or its characteristics and may disagree with its applicability to a particular group of women in a particular country. A close look will reveal that marianismo is a concept that is seriously flawed. In fact it is an extrapolation from impressionistic data that has been used mistakenly to account for the gender arrangements of an entire continent. The critique presented here echoes the stance adopted by the editors of this volume: Feminist scholarship should be grounded in the cultural, geographic, and historical specificity of gender arrangements. We must "desalambrar" the theoretical frameworks that have been imposed on the study of women and gender in Latin America.

SCHOLARLY RESPONSES

One of the first scholars to challenge Stevens, historian Silvia Marina Arrom (1980) warned that Stevens's ideas resembled the Victorian "cult of womanhood" found in the United States and Great Britain. However, noted Arrom, more recent scholarship tended to contradict the passive, powerless, self-sacrificing, and dependent women described by Stevens. Furthermore, new research contradicted her claim that marianismo had existed on the continent since the beginning of Spanish colonization. Though she did not dismiss the concept altogether, Arrom's data "strongly [suggested] that *marianismo* was not in fact a deep-seated Latin American cultural trait, but merely a variant of Victorianism introduced in the second half of the nineteenth century" (1980, 262).

Susan C. Bourque (political scientist) and Kay Barbara Warren (anthropologist) rejected the idea of separate spheres as well as the specific arguments offered by Stevens, arguing that, "because they are based on images rather than actual examinations of the politics of family life," they do not provide precise information on the power that the women exert, its material base for example; and they are "largely directed toward urban society and segments of the middle and upper classes in which wives do not work outside the home" (1981, 61). In contrast, anthropologist Vanda Moraes-Gorecki did use marianismo in her research, though she criticized Stevens's simplistic claim that machismo was unconnected with other forms of political subordination, her lack of class analysis and the uncritical acceptance "of folk images and middle class preconceived stereotypes of male/female relations" (1988, 28), and her conclusions that tended to justify the status quo on issues of sex domination and class oppression. Though Moraes-Gorecki accepted the basic idea of marianismo, she did not agree that it could be found all over Latin America and across all social classes. She showed that women had resisted domination in Latin America from the sixteenth to the eighteenth century, when privileged women had exercised power despite conventions, as well as in the early to mid-twentieth century when they fought for the vote in Argentina, Uruguay, and Chile.

Ethnographers Carol Browner and Ellen Lewin found that scholars generally depicted women's roles as timeless, static, and unconnected with the varied socioeconomic conditions of their lives. In their view, Stevens's description of women as altruistic, selfless, passive, and morally pure failed to take into consideration the material basis of their behavior (1982, 63). Swedish anthropologist Kristina Bohman, responding to Stevens's idea that Latin American women are not oppressed but rather enjoy privileges because of the importance given to motherhood, agreed that among the women of a Colombian poor urban neighborhood, there might be "a potential for female influence and power." However, she noted that "[Stevens's] argument needs to be seen in the light both of class differentiation and of the patriarchal structure of the family." Her findings indicated that, at least among the urban poor, "women's lack of economic resources and of opportunities for independent action severely curtails their possibilities of asserting their interests in any way at all" (1984, 35–36).

In her 1992 dissertation, anthropologist Marit Melhuus asserted that "For anyone working on Latin American gender ideologies it is not possible to overlook the notions of *machismo* and *marianismo*." However, Melhuus did not find Stevens's binary concept useful, noting that because it did not indicate how its complementarity is constructed, or "the weight of either with respect to the other," it ignored intrinsic aspects of the relationship between its parts. Furthermore, Melhuus and co-author Stølen later asserted that, "Not only did Stevens make generalizations about gender relations based on limited empirical material from middle class Mexico, she also neglected the importance of economic and political conditions for the constitution of gender relations" (1996, 12).

Fieldwork in two Guatemalan communities underpinned anthropologist Tracy Bachrach Ehlers's stance that while "marianismo appears to ring true because it provides a rationale for female subordination and because many women respond to it," it "idealizes the harsh reality of women's lives" (1991, 13). Ehlers's paper exemplifies the legitimacy attained by Stevens's concept, notwithstanding the persistent criticisms leveled against it. She had little use for marianismo, yet she found it necessary to position her own work in reference to the universe carved out by Stevens because she believed that in the U.S. academy marianismo "has evolved into a nearly universal model of the behavior of Latin American women" (1991, 1). In sum, a number of scholars writing on gender roles in Latin America, including anthropologists, seemed compelled to mention marianismo, rejecting it totally or accepting it in part, even when it may not have fit the specific issue and/or country they were studying.

Scholars based in Latin America have remained relatively uninterested in the concept. Among the exceptions is the Chilean scholar Sonia Montecino for whom marianismo is a foundational Latin American narrative, a foundational myth "that solves our problem of origin," in that "... it is a fundamental anchor of the mestizo imagination, of its culture more connected to ritual than the word" (1991, 28). Marianismo as an ideology of power is found in Zaira Ary's research project on masculinity and femininity in the Brazilian Catholic Church (1990), and is foregrounded in Alicia Del Campo's (1987) exploration of how poor Chilean women appropriated the most traditional images about them in a structurally symbolic confrontation with the fascist state at a time when the military used the same images to attempt to transform Chile's female population into passive subjects. Marianismo, as a space of ideological struggle, could acquire a liberating and revolutionary value when used effectively by women.

Though Peruvian anthropologist Norma Fuller (1992) accepted Stevens's basic argument, she had several reservations: Marianismo is a product of a modern *mentalité* and therefore a recent phenomenon; the separation between the public and the private sphere is not always clear cut; family ties still play an important role in Latin American politics; and the division of man (sexed) and woman (unsexed) does not explain per se the difference of power between men and women in the public arena. Finally, in an article first published in the early eighties, Mexican historian Josefina Zoraida Vázquez dismissed marianismo as a distortion and an exaggeration (1994, 18).

STEVENS'S ARGUMENTS

The most frequently mentioned article by Stevens and the one usually criticized is *"Marianismo:* The Other Face of *Machismo."* She first presented it at the 1971 Latin American Studies Association international congress and later published it in *Female and Male in Latin America,* edited by Ann Pescatello (1973a). The same year Stevens published a shorter version for the general public (1973c). In order to follow the full development of Stevens's ideas, two other sources must also be taken into account: "The Prospects for a Women's Liberation Movement in Latin America" (1973b) and "Mexican *Machismo:* Politics and Value Orientation" (1965).

Stevens's most important article is still *"Marianismo:* The Other Face of *Machismo."* It is here that she explained marianismo as a "pattern of attitudes and behavior" that has a counterpart in machismo, "the cult of virility" (1973a). Machismo is the mirror image of marianismo, or, as the title of her article states unmistakably, it is the other face of marianismo. They are twin phenomena, they complement each other perfectly and neither can exist without the other.

Other than stating that the main characteristics of machismo are "exaggerated aggressiveness and intransigence in male-to-male interpersonal relationships and arrogance and sexual aggression in male-to-female relationships" (1973a), Stevens did not discuss machismo at great length in "Marianismo" because she had supposedly examined that concept in her 1965 article entitled "Mexican *Machismo:* Politics and Value Orientations." In that text, she noted that there was little written about machismo from a social science perspective beyond Oscar Lewis's *Five Families;* J. Mayone Stycos's study of family and fertility in Puerto Rico; Samuel Ramos's 1934 study *El perfil del hombre y la cultura en México;* two films (*El Sietemachos* and *Animas Trujano*); Octavio Paz's ubiquitous, brilliant, impressionistic, and highly misogynist *The Labyrinth of Solitude;* and a handful of other sources. Nevertheless, she believed that all Latin Americans somehow "knew" about machismo since "it is all about them and it is as familiar to them as the air they breathe" and foreign observers include "its many manifestations in a list of distinguishing characteristics about the region" (1965, 848). At times her comments refer to Latin American males; but her sources, such as they are, deal only with Mexico (except for J. Mayone Stycos) and she repeatedly speaks of Mexican men. She explained, for example, that the latter equate virility, "a cultural symbol," with "exaggerated aggressiveness and with intransigence." When threatened, "a Mexican male will often react by shouting, 'I am a *macho*' (gesticulating toward his external genitalia), and will impute feminine characteristics to his adversary" (1965, 848).

Stevens did not seem concerned about the lack of social science research on a topic she considered so important for her argument. Her discussion of machismo was short, descriptive, superficial, and trite. However, she went on to address the possible connection between what she understood to be the two distinguishing characteristics of machismo, intransigence and violence, and the Mexican political system, a "sphere of activity peculiarly appropriate to males" (1965, 849). This last comment is quite odd, considering that in 1965 politics was an "activity peculiarly appropriate to males" not only in Mexico but also in much of the world, including of course the United States.

"*Marianismo*" has a short introduction followed by three parts and a conclusion. The first section, "Old World Antecedents" explores what Stevens calls the ancient roots of marianismo in Old World cultures. Here she presents a historical overview of the cult of the Virgin Mary and the worship of ancient goddesses. The second section entitled "New World Development" contains a discussion of marianismo in contemporary Latin America. The third section, "Alternative Models," includes a few generalizations, the mention of exceptions, and comments about the power of marianismo.

Stevens presents her discussion of marianismo and its relationship with machismo in her introductory remarks. She begins by establishing that every society has a "pattern of expectations based on real or imagined attributes of the individuals or groups who perform certain tasks. With time, these attributes attain a validity which makes it possible to use them as criteria for value judgments quite unrelated to functional necessity" (1973a, 90). Having thus set the stage, she then defines and describes machismo in a short paragraph, repeating the exact wording she used in her 1965 article. Furthermore, she only provides one footnote—citing her own

1965 article—as if nothing else had been published since then or assuming her work was a sufficient reference.

Stevens indicates in a rather indirect form that she has learned about marianismo in Latin America but does not specify from whom. Presumably her informants were women, though she remarks that women generally do not speak about marianismo, possibly because many of them fear losing the power they draw from it. However, in a curious paragraph she adds that in recent years some men have begun to write about it and one has called it "*hembrismo*" or "female-ism" and another one "*feminismo*" (1965, 849):

> It has only been in the quite recent past that any attention has been focused on the other face of the problem [marianismo]. Women have generally maintained a discreet reserve with respect to the subject of marianismo, possibly because a very large segment of the group fears that publicity would endanger their prerogatives. A short time ago, however, a handful of male writers began to focus on this heretofore neglected pattern of attitudes and behavior. In this way, the term *hembrismo* ("female-ism") has been introduced by one observer, while *feminismo* has been used by another (1973a, 849).

A footnote indicates that the source for these terms is a 1970 special issue of *Mundo Nuevo*—a famous cold war journal that appeared in Paris—with six articles on what we presently would call gender roles in Latin America.

Building on the ideas in this special issue, Stevens states that she found the terms "hembrismo" and "feminismo" unsatisfactory, preferring "marianismo" (1973a, 100). The implication here is that hembrismo, feminismo, and marianismo are equivalent concepts. In fact, the three words have very different meanings. Hembrismo was an old word in Mexico, first used to describe women's passivity as a complement to the machismo of Mexican men by the Mexican scholar María Elvira Bermúdez (1955), for whom hembrismo entailed exaggerated feminine characteristics—weakness, acquiescence, and inertia; it *was* the other face of machismo and it spelled out powerlessness and dependency, as it did for the scholar Fernando Peñalosa and one of the authors of the *Mundo Nuevo* articles, Salvador Reyes Nevares.[2] Bermúdez also saw religious connotations with the Virgin Mary. Hembrismo as it was generally understood in Mexico, was therefore very different from marianismo as Stevens defined it because for her marianismo was *the* source of women's power. Similarly, marianismo in no way could be equated with feminismo. In the *Mundo Nuevo* article, feminismo meant a struggle for women's liberation, a far cry from marianismo as Stevens defines it in her text.

Because they are cited by Stevens as a basis for her ideas, it is important to examine the content of the *Nuevo Mundo* articles further. Mexican writer

Salvador Reyes Nevares discusses machismo and its complementary phenomenon hembrismo, noting that machismo is not something that defines all Mexicans, only some of them. If a machista is aggressive, a heavy drinker, and he needs to affirm his own importance it is in relation to a specific woman; as his complement, she will be passive, patient, faithful, sweet, and so on. Out of this dynamic, Mexicans have created the commercial image *machihembrismo,* the double image of the macho man and the passive woman, sung about in ballads and depicted in movies, an image which does not really exist in Mexico but unfortunately represents Mexicans throughout the world.[3] For Sebastián Romero Buj, machismo in Latin America is only a fiction to be found in literary texts.[4] Peruvian journalist José B. Adolph gives a tongue and cheek manifesto for the liberation of women or feminism, which he prefers to call hembrismo, thus giving a new meaning to the term.

In the same special issue, journalist and feminist Ana María Portugal took stock of the recent changes in the lives of middle-class women in the Peruvian capital. Using letters published in women's magazines and surveys, she offered not only personal reactions but also specific information about the growing presence of women in higher education. Women studied because they wanted to be independent: "Lima is no longer afraid of the steady growing wave of women committed to conquer man's world, through universities and offices" (1970, 23). Furthermore, despite the repressive, machista, and hypocritical nature of Peruvian society, women were beginning to be sexually freer. In other articles, Haydée M. Joffre Barroso underscored dramatic changes in Argentine women's lives, and two Chilean journalists, Rosa Cruchaga de Walker and Lilian Calm reported the answers to a series of questions posed to six professional women. These sources provided information contradicting the thesis advanced by Stevens; not surprisingly, she did not discuss them. When she did use the information they contained, for example the women described by Portugal, it was to present these women as exceptions to the rule of marianismo. Although the articles focused on concepts that concerned her and one wrote about the same binary opposition she used, once again she chose not to comment on any of them.

Stevens asserted that marianismo and machismo are New World phenomena. She insisted that "the fully developed syndrome occurs only in Latin America" (1973a, 91) as a consequence of the Spanish conquest. Citing the Spaniard Julio Caro Baroja, she saw machismo as a degeneration of the concepts of honor and shame associated with the notion of manliness brought by conquistadores and adventurers. Unfortunately, she does not discuss why the New World environment was "propitious" for the development of or how the differences found by the Spaniards (she does not mention the Portuguese) in the various areas of the continent influenced machismo. Nor does she explain the connection between what "the soldiers and adventurers" brought in the sixteenth century and what she sees in the twentieth. One is left to believe that what she calls contemporary machismo and marianismo are exactly what they were in the sixteenth century.

Considering her discussion of machismo sufficient, Stevens then turns to marianismo without explaining how, when, or why the connection between the two concepts was established. She finds that "all mestizo social classes are permeated with machismo and marianismo characteristics" (1973a, 91) while indigenous communities are free from these patterns of behavior as long as they retain what she calls "their cultural purity." This is another odd statement in view of the extended and extensive presence of Europeans in the Americas. Furthermore, she does not explain what "mestizo social classes" are, in what country or countries they are to be found, and offers neither examples nor sources to clarify this statement.

Stevens then proceeds to offer an etymological discussion of marianism and marianismo that is highly problematic because the word marianism does not exist in English and marianismo as she uses it, "a secular edifice of beliefs and practices related to the position of women in society" (1973a, 91), does not exist in Spanish. Marianismo in Spanish refers without equivocation to the cult of or devotion to the Virgin Mary. Stevens created marianismo as an analytic category. In her article "The Prospects for a Women's Liberation Movement in Latin America," she confessed that marianismo "is a term that I have coined but which I hope will be adopted in future discussions of the subject, because it reflects the present complex psychic content and connotes the historical development of the pattern" (1973b, 315).

Nevertheless, in "*Marianismo,*" she traced the historical and prehistoric roots back to the Mediterranean world and ancient Middle Eastern societies where marianismo supposedly originated in the "primitive awe at woman's ability to produce a live human creature from inside her own body" (1973a, 92) as manifested in the form of goddess worship in Mesopotamia and other areas of the Mediterranean Sea. She pinpointed the origin of the Marian cult in the western hemisphere with Our Lady of Guadalupe; in 1810, Father Hidalgo raised her banner when he led a revolt against the Crown, and 100 years later, Pope Pius X declared her the patroness of Latin America. Though Stevens considers this sequence of events sufficient to establish the importance of the cult of the Virgin in Latin America, she never mentions the symbolic transformations of the Virgin throughout the centuries, and she ignores the impact of the Church as an institution. She admits that she is unable to explain a crucial point in her argument, namely "how the excessive veneration of women became a distinguishing feature of Latin America." Still, Stevens states, "Latin American mestizo cultures— from the Rio Grande to the Tierra del Fuego—exhibit a well-defined pattern of beliefs and behavior centered on popular acceptance of a stereotype of the ideal woman ubiquitous in every social class. There is near universal agreement on what a "real woman" is like and how she should act" (1973, 92). Stevens would have us believe that, regardless of class origins, cultural specificity, religious beliefs, political convictions, ethnic origin, educational levels, or history, the ideal Latin American woman is everywhere the same, semidivine, morally superior, and spiritually strong, and everywhere she behaves according to this ideal.

Stevens tells us that very early in life Latin American women acquire an infinite capacity for humility and sacrifice, an endless patience, and a deep sense of obedience and submission to the demands and desires of fathers, brothers, sons, and especially husbands, who are obstinate, intemperate, and foolish, like children who must be humored. A Latin American woman is "complaisant with her mother and her mother-in-law because they are supposedly reincarnations of the great mother" (1973a, 95). She is a sad creature, who suffers because of men's sins and is condemned to mourn during long periods for an endless number of relatives, while the men escape to enjoy life with their mistresses. She is a virgin when she marries and frigid for the rest of her life, because only "bad women" are supposed to enjoy sex. " 'Good' women do not enjoy coitus; they endure it when the duties of matrimony require it" (1973a, 96).

Stevens admits that there are exceptions, that "not all Latin American women conform to the stereotype prescribed by *marianismo*." There are also "bad women" who are not seen as "real women" and among middle- and upper-class women "norms of sexual behavior are often disregarded in practice." Nevertheless, "the image of the black-clad mantilla-draped figure, kneeling before the altar, rosary in hand praying for the souls of her sinful men-folk, dominates the television and cinema screens, the radio programs, and the popular literature, as well as the oral tradition of the whole culture area" (1973a, 96).

It is not always clear when Stevens refers to behavior, stereotypes, or images, but in the last analysis, she argues that marianismo does dictate the behavior of Latin American women. "Some women choose to pattern their behavior after the mythical and religious ideal symbolized by the figure of the Virgin Mary. Others deviate from this ideal to greater or lesser degree in order to obtain the satisfaction of their individual desires or aspirations. The ideal itself is a security blanket which covers all women, giving them a strong sense of identity and historical continuity" (1973a, 98).

The representations of women depicted by Stevens in the 1970s may indeed have been found in some countries, in movies, novels, and endless *fotonovelas* and *radionovelas* of the sixties and the seventies, as they may be in the *telenovelas* of today. Some women from several countries might recognize themselves in her descriptions or come close to them; but the lives of Latin American women even in 1973 were far more varied and complex than her single dominant stereotype. In fact, as she was writing, women in many countries were swelling the ranks of higher education in growing numbers, entering new professions, founding feminist collectives, staying in the labor force (even after marriage), joining guerrilla organizations, experimenting with sex before marriage, having illegal abortions, and if they lived in countries where they could not divorce, they separated anyway and remarried in Uruguay or Mexico or just did not get married again. And then, of course, there were the women who were single heads of household, who had always done wage work, cleaning, washing, ironing, or worked long hours in the fields, in sweatshops or factories, in classrooms, in front of a telephone board, a typewriter, a sales counter, or in a hospital.

Stevens warns her readers that, while English-speaking feminists might see the hand of tyrannical men in the condition of Latin American women and occasional Latin American women might find their situation repressive, "it is quite apparent that many women contribute to the perpetuation of the myths which sustain the patterns described." Why would they work against their own interests—if indeed, they do? Might it not be possible that while employing a distinctive repertory of attitudes, they are as "liberated" as most of them wish to be (1973a, 99–100)? She admits that sometimes the pressures to conform were too harsh on certain women, and she offers the examples of Sor Juana Inés de la Cruz and Manuela—no last name, just "the mistress of Simón Bolívar." But for her, the exceptions only confirmed the rule: Women in Latin America have options; they can follow the dictates of religion and myth or pursue "an earthy and hedonistic lifestyle" or "an achievement-oriented puritan ethic." But to all, marianismo is the "security blanket which covers all women." (1973a, 98).

Largely on the basis of images, impressions, and personal observations seemingly centered on Mexico and Puerto Rico, Stevens suggested that marianismo—throughout the continent—governs the lives of Latin American women almost exclusively by its unwritten rules. She wrote as if there were no economic forces at work, political parties had never excluded women from their ranks (except as auxiliary groups), women's civil rights had never been curtailed, rural women lived in the same conditions as urban women, and the Catholic Church had never exerted power and influence over women's sexuality or had nothing to do with the role of marriage in society. She did not see any connection between the absence of divorce legislation and the influence of the Catholic Church. For her, the "fiction of unassailable purity conferred by the myth on saint and sinner alike makes divorce on any grounds a rather unlikely possibility" (1973a, 98). Ignoring the history of social reform in many countries, she overlooked the changes in the Church and its role or that of Left parties in adopting protective legislation for working mothers at the turn of the century. Instead, she saw once again the role of marianismo: "The granting of sick leave to the mother of a sick child is not so much a matter of women's rights as a matter of the employer's duty to respect the sacredness of motherhood which the individual woman shares with the Virgin Mary and with the great goddess of pre-Christian times" (1973a, 99). Only when assessing the impact of marianismo on labor force participation was she more cautious, admitting that there was not sufficient information, though "we must leave the possibility that a considerable number may have freely chosen their *marianismo* cake and eat it too" (1973a, 99).

It would seem that in this instance, Stevens chose once again to disregard available research or just did not look for it. She could have used Nora Scott Kinzer's "Women Professionals in Buenos Aires," (1973) and "Priests, Machos and Babies" or "Latin American Women and the Manichaean Heresy" (1973). In the case of Mexico, there was already information about the diversity of family patterns (and the predominance of single-headed households among poor women, something Stevens does not mention at all), and there was data about the participation of women in the labor force as well as their level of education.[5]

In her conclusions, Stevens noted that "mestizo societies" are changing and there are signs that younger generations, in particular middle-class university students, have "somewhat" different values than their parents, noticeable during the 1968 Mexico student strike. However, she did not believe that marianismo was destined to disappear for a long time, because she did not expect Latin American women to vote as a bloc to change divorce laws or to abolish sex discrimination; "they are not ready to relinquish their female chauvinism" (1973a, 100).

In her 1973 article on the limited prospects for women's liberation in Latin America, Stevens reiterated her lack of faith in Latin American women—who in several countries, together with men, changed divorce laws and much more in subsequent decades. She claimed that while the social and political conditions existing in "North-Atlantic post-industrial societies" contributed to the emergence of a women's movement, Latin American relations between the sexes were not secular in nature but were "suffused with sacred significance" (1973b, 315). She then proceeded to explain the "*machismo–marianismo* pattern" of attitudes and behavior, which she saw as a variant of the social division of labor. Citing Portugal, she mentioned that there are exceptions and declared that the "myth of Latin American male dominance is just that: a myth" perpetuated by women in part "because it preserves inviolate a way of life which offers many advantages for them" (1973b, 316). She went on to examine the power of machismo–marianismo in what she characterized as mestizo countries (Mexico, Peru, Ecuador, Brazil) and countries of mostly European population (Chile, Argentina, Uruguay). She concluded that because of the benefits that women accrue from marianismo (with so many low-paid domestic servants, the "monotonous and demeaning aspects of household work are unknown to them as have been the burdens and restrictions of child care" [1973b, 317]), and as long as the indigenous populations remained marginal to modernization and industrialization, despite the signs of economic transformations and an occasional lonely and "rather rare" voice of protest like Portugal's, she saw little chance for the emergence of a women's liberation movement among middle-class Latin American women.

But Portugal was not alone in embracing feminism. In fact, precisely in 1973, she heard about Alimuper (Acción para la liberación de la mujer peruana), the first Peruvian feminist group, which was founded that year and organized a demonstration in Lima to protest beauty contests; she joined it and has been a feminist ever since. In Buenos Aires, Argentina, there were already three *colectivos feministas*. In 1970, a small group of women—including, among others, María Luisa Bemberg, who would soon begin to direct films, the well-known photographer Alicia D'Amico, and the writers Leonor Calvera and Leonor Cano—began to meet in a consciousness-raising group and founded UFA (Unión Feminista Argentina and a colloquial expression meaning "enough!"). A second group emerged also in 1970. *Nueva Mujer* had a short life but managed to publish two books: *Las mujeres dicen basta* and *La mitología de la femenidad*. These groups were joined in 1972 by the MLF (Movimiento de Liberación Femenina), founded by María Elena Oddone. The Mexican feminist movement was also launched in 1971 with the founding of MAS (Mujeres en Acción Solidaria), a group that came together after the journalist Marta Acevedo, on September 30, 1970, published a long article about the feminist movement in the United States in the supplement of the magazine *Siempre!*, "Nuestro Sueño está en escarpado lugar (crónica de un miércoles santo entre las mujeres), 'Women's Liberation.' San Francisco" (Our Dream is in a steep place [Chronicle of a holy Wednesday among women], "Women's Liberation." San Francisco). They organized their first demonstration on Mother's Day, May 9, 1971. Thus, the Latin American second wave women's movement was well under way precisely when Stevens and other U.S. scholars were writing that there was little chance for it to exist.

CONCLUSIONS

Stevens's work on Latin American women was not grounded in systematic original research and did not rely on then-available research carried out by other scholars. She made cursory use of a limited number of sources, ignored history, overlooked profound differences in the region, and when the available research contradicted her views, she did not mention it. What she did was to use the old (machismo/hembrismo) Mexican formula—a superficial set of images contested by many Mexicans but prevalent in popular culture—repackage it, give it a new name (marianismo), elaborate a supposedly scholarly rationale for it, and claim that it existed throughout Latin America.

In her defense, it could be argued that when Stevens wrote on Latin American women it was a new field in which few scholars were involved. When Stevens presented her paper at the Latin American Studies Association congress, it was the first time that a panel entirely composed of women presented papers on women. The problem is that, despite its serious flaws, almost 30 years later we still find scholarship in which marianismo is presented as a serious conceptual framework; when it is dismissed because of its limitations or inaccuracies, it is nevertheless mentioned as a referent. In an anthology reprinting Stevens's article, historian Gertrude M. Yeager stated that "*Marianismo* has been the organizing tenet of Latin American women's studies in this country since the appearance of Evelyn Stevens' essay, which examines gender (learned), not sex (biological), roles in terms of *machismo* and *marianismo*" (1994, 3). Although this is surely an exaggeration, marianismo nevertheless acquired a legitimacy it did not merit.

Why, then, did marianismo gain legitimacy among U.S. academics? And how do we explain its longevity? These questions are not central to my task, but I must confess that I am intrigued by them. One explanation may be that research on gender roles or more generally the ideological dimensions of women's subordination was slow to develop among U.S. Latin Americanists. When scholars, especially anthropologists, began to do research in these areas in the late seventies and early eighties, they found one basic article: Stevens's essay. While most found its content inadequate and proceeded to demonstrate their point with their own research, they nevertheless were caught in its trap. Though we find fewer examples today of anthropologists using the concept, anthropology's role in the seventies and eighties in legitimizing this framework fed a specter that continues to haunt interdisciplinary feminist scholarship on Latin America.

In response to the challenge posed by the editors of this volume, I would say that "*el alambre*," or the wire, comes in many forms. Marianismo is an ahistorical, essentialist, anachronistic, sexist, and orientalist fabrication. Marianismo has been a highly prickly and particularly sterile barbed wire that has fenced in research on women and gender in Latin America for too long. It is high time for it to be cut and set aside or buried deep in the ground, for once again, the emperor—in this case the empress—has no clothes.

NOTES

In writing this essay I was fortunate to count on an excellent senior research paper by my student Melanie Watts, Dartmouth, class of 1999. Her help was invaluable and I am deeply indebted to her. I would also like to thank Patricia Carter, Marianne Hraibi, and Miguel Valladares from Baker Library, for their help, their infinite patience, and their good humor. Thank you also to my colleagues and friends Susan C. Bourque and Gene R. Garthwaite for their useful comments and good advice, and to my friend and compañera, Ana María Portugal, for her quick and helpful answers to my questions.

1. Anthropologist Mary Elmendorf (1977) was one of the first scholars to mention *marianismo*.
2. See also Peñalosa 1968 and López-Capestany 1973.
3. Américo Paredes 1967 makes a similar argument; however, he also explains the connection between specific historical conditions and the emergence of machismo both in the United States and Mexico.
4. Argentine sociologist Julio Mafud's (1967) references on machismo were also mostly literary.
5. Elú de Leñero (1971), Hernández (1971), Leñero Otero (1968), and Ruddle and Odermann (1972).

REFERENCES

Adolph, José, B. 1970. "La emancipación masculine en Lima." *Mundo Nuevo* 46: 39–41.

Arrom, Marina. 1980. "Teaching the History of Hispanic Women." *The History Teacher* 13(4): 403–506.

———. 1985. *The Women of Mexico City, 1790–1857*. Stanford: Stanford University Press.

Ary, Zaira. 1990. "El *marianismo* como 'culto' de la superioridad espiritual de la mujer." In *Simbólica de la feminidad. La mujer en el imaginario mítico-religioso de las sociedades indias y mestizas*, edited by Milagros Palma. México: MLAL and Ediciones ABYA-YALA.

Bartra, Roger. 1987. *La Jaula de la melacolía.* México: Editorial Grijalbo.
Bermúdez, María Elvira. 1955. *La vida familiar del mexicano.* México: Antigua Librería Robredo.
Bohman, Kristina. 1984. *Women of the Barrio: Class and Gender in a Colombian City.* Stockholm: University of Stockholm.
Boserup, Ester. 1970. *Woman's Role in Economic Development.* London: George Allen and Unwin.
Bourque, Susan C. and Kay Barbara Warren. 1981. *Women of the Andes: Patriarchy and Social Change in Two Peruvian Towns.* Ann Arbor: The University of Michigan Press.
Browner, Carol and Ellen Lewin. 1982. "Female Altruism Reconsidered: The Virgin Mary as Economic Woman." *American Ethnologist* 9(1): 61–75.
Del Campo, Alicia. 1987. "Resignificación del *marianismo* por los movimientos de mujeres de oposición en Chile." In *Poética de la población marginal: sensibilidades determinantes,* edited by James V. Romano. Minneapolis, MN: The Prisma Institute.
Ehlers, Tracy Bachrach. 1991. "Debunking *marianismo:* Economic Vulnerability and Survival Strategies Among Guatemalan Wives." *Ethnology* 30(1): 1–16. (1993. Reprinted in *The Other Fifty Percent: Multicultural Perspectives in Gender Relations,* edited by Mari Womack and Judith Marti. Prospects Heights, IL: Waveland.)
Elmendorf, Mary. 1977. "Mexico: The Many Worlds of Women." In *Women, Roles and Status in Eight Countries,* edited by Janet Zollinger Giele and Audrey Chapman Smock. New York: John Wiley and Sons.
Elú de Leñero, María del Carmen. 1971. *Mujeres que hablan.* México: Instituto Mexicano de Estudios Sociales.
Fuller, Norma. 1992. "En torno a la polaridad machismo–marianismo." *HOJAS DE WARMI* 4 (November): 5–10.
Hernández, Susana. 1971. "Unas características de la mujer mexicana de clase media." *Revista Mexicana de Ciencia Política* (July–September).
Kinzer, Nora Scott. 1973. "Priests, Machos, and Babies: Or, Latin American Women and the Manichean Heresy." *Journal of Marriage and the Family* 35(2): 300–312 and "Women Professionals in Buenos Aires." *Female and Male in Latin America,* edited by Ann Pescatello, 159–190. Pittsburgh, PA: Pittsburgh University Press.
Leñero Otero, Luis. 1968. *Investigación de la familia en México.* México: Galve.
———. 1971. "The Mexican Urbanization Process and Its Implications." *Demography* 2: 866–873.
López-Capestany, Pablo. 1973. "Exploración del *machismo:* Particular referencia a Gabriel García Márquez." *Cuadernos Americanos* 189(32): 105–121.
Mafud, Julio. 1967. "El machismo en la Argentina." *Mundo Nuevo* 16: 72–78.
Melhuus, Marit. 1992. " 'Todos tenemos madre: Dios también' Morality, Meaning and Change in a Mexican Context." Ph.D. diss. University of Oslo.
Melhuus, Marit and Kristi Anne Stølen. 1996. *Machos, Mistresses, Madonnas: Contesting the Power of Latin American Gender Imagery.* London: Verso.

272   *Marysa Navarro*

Moraes-Gorecki, Vanda. 1988. "Cultural Variations on Gender: Latin American *Marianismo/Machismo* in Australia." *Mankind* 18(1): 26–35.

Montecino, Sonia. 1991. *Madres y huachos. Alegorías del mestizaje chileno.* Santiago: Editorial Cuarto Propio. CEDEM.

Montecino, Sonia, Mariluz Dussuel, and Angélica Wilson. 1988. "Identidad femenina y modelo mariano en Chile." In *Mundo de mujer: continuidad y cambio.* Santiago de Chile, Ediciones Cem.

Paredes, Américo. 1967. "Estados Unidos, México y el machismo." *Journal of Inter-American Studies* 9(1): 65–84.

Peñalosa, Fernando. 1968. "Mexican family roles." *Journal of Marriage and the Family* 30(4): 680–789.

Portugal, Ana María. 1970. "La peruana ¿'Tapada' sin manto?" *Mundo Nuevo* 46: 20–27.

Reyes Nevares, Salvador. 1970. "El machismo en México." *Mundo Nuevo* 46: 14–19.

Romero Buj, Sebastián. 1970. "Hispanoamérica y el machismo." *Mundo Nuevo* 46: 28–32.

Ruddle, Kenneth and Donald Odermann. 1972. *Statistical Abstracts of Latin America.* Los Angeles: University of California Press.

Stevens, Evelyn, P. 1965. "Mexican *Machismo,* Politics and Value Orientations." *The Western Political Quarterly* 18(4): 848–857.

——. 1973a. "*Marianismo:* The Other Face of *Machismo.*" *Female and Male in Latin America,* edited by Ann Pescatello, 89–101. Pittsburgh, PA: Pittsburgh University Press.

——. 1973b. "The Prospects for a Women's Liberation Movement in Latin America." *Journal of Marriage and the Family* 35(2): 313–321.

——. 1973c "*Machismo* and *Marianismo.*" *Society* 10(6): 57–63.

Vázquez, Josefina Zoraida. 1994. "Women's Liberation in Latin America: Towards a History of the Present." In *Confronting Change, Challenging Tradition: Women in Latin American History,* edited by Gertrude M. Yeager. Wilmington, DE: Scholarly Resources, Inc.

Womack, Mari and Judith Marti, eds., 1993. *The Other Fifty Percent: Multicultural Perspectives in Gender Relations.* Prospect Heights, IL: Waveland.

Yeager, Gertrude M., ed. 1994. *Confronting Change, Challenging Tradition: Women in Latin American History.* Wilmington, DE: Scholarly Resources, Inc.

13. Understanding Gender in Latin America ᕽ

Sonia Montecino
(Translated by Deborah Cohen and Lessie Jo Frazier)

My commentary poses a question about Latin American mestizo culture that I have been trying to formulate and adjust for years in various texts.[1] Here I simply want to add a new twist: To bring the intellectual production of Latin American women and a discourse about the feminine subject together with theoretical underpinnings from dominant territories (Europe and the United States).

The issue of Woman and, later, of gender has been present in our continent since the decade of the 1970s, conditioned by knowledge coming from the developed world. At an important moment Woman, as a universal category, was a favorable field for generating a syntax to explain the subordination and the resulting negation of Woman as constitutive elements in and of all societies. The notion of patriarchy associated with the category Woman, then, was the bedrock of a current of thought that saw that, with surprising ubiquity, everywhere the same motive operated and resulted in the devalued position of women. This lens brought with it numerous consequences. One of them, of great importance in the "first world," was that researchers embarked upon a systematic study of those negations (lacks), deconstructing existing ethnographic information. Another consequence of this early phase was the ghettoization of women's studies and the discovery that anthropology, as a discipline, was talking and thinking about non-Western societies from both a masculine and an ethnocentric position.

The anthropology of gender, which superseded the anthropology of women, was supposed to bring about a different way of looking at things: Woman, as a universal category, no longer would have the power to displace from view notions of gender and difference nor the specificity through which relations between the feminine and the masculine are constituted in particular cases. From there have emerged tendencies that emphasize the symbolic or social construction of gender differences. The prolonged debate about the construction of binary oppositions such as nature/culture and public/private gave rise to a never-ending series of investigations and analyses. At the same time, the position of the social construction of differences linked Marxist analyses with the feminine problematic and expanded anthropological studies about this issue in the fields of economics and political science. The last few years have seen the demand for a feminist anthropology that tries to overcome the limits of previous currents of thought to tackle the issue of gender, interweaving with it the categories of ethnicity and class, and including history as an explanatory element of difference.

This brief panorama of the development of ideas in the first world returns us to the question of whether these ideas have or have not experienced a (dis)tortion in Latin America. I say tortion (*torción*) or twist since I am thinking of Latin American subjects as a mixture, as mestizos; as such we have always struggled among ourselves between reason (*el logos*) and myth (Arguedas 1978, 1983) as the most pristine expressions of this phenomenon. At first glance we can say that the dynamics of thought about Woman appears in our territory as a copy, a translation of theoretical frameworks from the first world, a noncritical application of the paradigms of patriarchy, subordination, and above all the concomitant themes imposed by the metropole (see the abundant works on women in the economy, employment, and work). Suspiciously, the international agencies that finance studies on women in Latin America are more interested in action than in investigation, and inasmuch as action is accepted, it is required to be about "urgent problems" as defined by dominant discourses. Related to this, one can also appreciate that the issue of Woman in Latin America is conditioned by a global market in issues—for example, when "the youth decade" emerges, studies shift toward young women, when it is "the environmental decade," that relationship becomes pertinent, and so on.

When, as a scholar in the South,[2] one reads an extensive bibliography from the first world and sees the arguments, debates, theories, and countertheories, one immediately thinks about what happens in our case: We internalize an argument or discussion that we haven't made or worked out. One notes that the tradition of reason existing in the first world has hardly been approximated in our circumstances, with an absence of new texts and as heirs of a tradition that has always been rooted in ceremony and orality.

However, both in spite of and because of the precariousness of material resources, there has emerged in Latin America a specific knowledge about the silenced image of women in diverse spaces, classes, and cultures. I refer to the birth of life histories, of testimonies, of oral expression, which—transformed into text—tried to bring to light and reveal feminine life. Then studies came along that, although without a clear program, tried to realize the specificity of gender(ed) experience and the constitution of gender through a mythic world, approaching a stance that, grounded in given cultural particularities, could dialogue with and even challenge the universality of paradigms regarding Woman (see Palma 1990; Vega 1986, 1990). However, only in the last few years has there emerged an attitude of doubt and contradiction, a self-inspection that looks for the reflection of difference in its own reflection, a move making audible the critique and the search for new models of understanding. The complexity of our reality, comprised of multiple sounds, requires and demands that intellectuals bring differences to life (*hacer carne*).

What the intellectual production of Latin American women hints at is precisely to make explicit what "by habit keeps you quiet," as Octavio Paz has said (1989, 16). Perhaps it is just those silences—the things not said in studies and reflections—that you would have to tackle as a first step: the lack of a history of gender in Latin America (as Victor Toledo [1993] asserts); the encapsulating of voices on the borders of the disciplines; the necessity of a sensibility that goes beyond a mere compilation of data that corroborates what has already been established as the truth; the overcoming of the tendency to focus our knowledge toward the demands of the market; and, along with those issues, the necessary gesture that specifies who speaks and what is said by the silences, by the prohibitions and denials, and—why not—by culture. Perhaps this movement of thought shows that the fusion (*lo mestizo*) of our identity shapes our emphasis on the subjects that we want to describe, to put on stage, to reveal, and therefore, that the ambiguity that characterizes our knowledge[3] might be more of value than disgrace.

What do I mean by mestizo thought? Certainly, in the first place it has to do with a semiotic system more than a genetic one, although, as Jorge Guzmán (1991) asserts, the semantic and somatic go hand-in-hand in our Latin American cultures where each of us, "lives in a signifying relation to our body" (21). As Guzmán indicates, another of our community's characteristics is a great resistance to the recognition of mestizaje, in spite of which in all of us there is a dialectical interaction between two or more distinct cultures. Thus, Latin American subjects simultaneously operate within the diverse cultural frameworks from which we are constituted, independent of our ethnic origin; and our mestizaje entails that, moreover, one of these cultures will be dominant and the other(s) dominated and scorned. For this reason, we circulate "interminably between the ... cultures and it is not possible to invoke one without, in some form, having the other, that which always has been structured as the opposite, concomitantly invoked" (10). In this sense, when I say mestizo thought I refer on the one hand, to a thought in tension due to this oscillation, these hierarchies, and the traditions that coexist in conflict within our ways of thinking.

On the other hand, I use the term mestizo thought in the same context in which I developed my concept of the bastard, or *huacho*.[4] In this sense, it is a thought that grants us "legitimate" filiation within the power of knowledge. Above all, Latin American and Chilean women's writing and reflection, in my opinion, remains suspended under the rein of the "not-tradition." Only recently has there been in place a literary tradition of women's writing in various fields recognized by the academic and public worlds. Therefore, this thought is huacho: it started out as "illegitimate" in the realm of masculine traditions enshrined within the lineage of accepted discourses. If we focus our gaze now beyond national borders, it is evident that Latin American men and women share the "illegitimacy" and the huacheraje of our thought in relation to the "Occidental logos" and that the circulation of our ideas occupies only a secondary place in the universe of dominant discourses.

So, let us return to the theme of mestizo thought and discuss it in relation to texts produced about women and gender in our continent. Authors such as Julieta Kirkwood (1987), Milagros Palma (1990), Marcela Lagarde (1990), and Marta Lamas (Chapter 11, this volume), among others,[5] construct their theories, their interpretations, using, twisting (*torciendo*), amalgamating ideas, without wedding themselves to any single rigid theoretical framework. It is a weaving that proceeds by intertwining concepts useful for reading the particularity of their reality and, in this sense, it is a form of desalambrar-ing, of tearing down received models and theories. I use the word wedded precisely in the sense of an institutionality that forges parentage, legitimacy. Thus, this refusal to utilize a single referent but to instead mix is a baroque and mestizo gesture.

To be sure, there are concessions, there are "embarrassments," there are tensions, there is genuflection. But alongside these there appear other illuminating cultural frameworks. That means, for example, that Milagros Palma proposes that the violence against women in Latin America is rooted in the opposition conqueror/conquered and in the lack of a resolution of our past, placing la Malinche[6] as a metaphor for the ambivalence of the feminine in our countries; or that Marcela Lagarde makes us obsessively revisit—mimetically with our culture—all of the possibilities that the mother has—body and sense—in our communities; or that Julieta Kirkwood confronts us with "history's knots" using the allegory of the tree rings—that is to say, as in the chronology of nature—in order to twist (*torcer*) or turn around the official periodizations of Chilean history. I cannot provide here a complete hermeneutic of mestizo thought in the work of Latin American intellectuals because this corpus has yet to be compiled and because here I am only interested in accenting a field that should be approached epistemologically from an "other" reading.

By scrutinizing and accepting our mestizo identity we can find our own particular "sound" that resembles our position from which our thinking proceeds. Making explicit the mixture of theories and models, and our fragmentary use of them, we can construct an "other" way of knowing; the same goes for valuing our own genres of work, such as literature, myths, and legends. Finally, a scrutinizing of ourselves involves inspecting the outside and inside, complete with their complex determinations and complications. Thus, our colloquialisms encompassed in silences can be transformed into a fertile ground from which we speak for ourselves (make theory) and that can be spoken about by others (tempered by theory) with whom we, without a doubt, have a relation of both tension and solidarity. The interplay between what is ours—our syntax—and the appropriated—the syntax that expresses us—can result in an atmosphere of reflection on gender and constitutes one possible path toward the construction of a fertile synthesis.

The last few years have been fruitful in the sense of deterritorializing, of globalizing. And it seems to me that the movement to rupture the borders has been positive for women and our discoveries made in this rupture nourish us. The networks that supposedly decenter the nuclei of hegemonies do work. Nevertheless, I sense, each time with greater certainty, that the power of knowledge gets reproduced in these networks and that those who are not integrated into them find their "truths" excluded from competition or from even being known. Thus deterritorializing brings with it its own fences.

Even so, I am optimistic that globalization has as one of its consequences the manifestation of diversities and identities. That is to say, the more homogenization—in consumption, in discourses, in knowledges—the greater the longing for singularization. More than ever the tension between the particular and the universal appears on the scene. But also more than ever, we find reproduced the mechanisms and the abyss between the few who possess much and the many who possess little. This is manifested in the possibilities that exist in Latin America for women to obtain economic support to think. This is why the competition in and between countries is patent; it is a competition evidenced in the omissions, in the not-cited, in the frequent making of clean slates of what others on the continent have produced. These negations have to do with the struggle for funding and also with the mestizo gesture of negating the names of others because they don't have "prestige" or legitimacy within the realm of those who must be seduced: the agencies, the foundations, the donating institutions, and so on.

To deterritorialize provides us with possibilities to generate more ideas, images, and hypotheses that oblige us to reread, to re-elaborate, that is to say, to make our mestizaje even more prolific. However, it also entails the risk of silencing us. I mention this because the power structures and the circulation of knowledge continue to be the prerogatives of the networks of women who have more access to resources or who know how to best "negotiate" or "move" their projects. Nevertheless, I am confident that it also brings with it the possibility of being included in an anthology such as this one, to cross frontiers in solidarity, to form part of a movement that is not interested in having market values—success, competition, individualization—become the values of politics and culture. Down the line we will see whether the gesture of desalambrar in the anthropology of gender in Latin America will have transformed "that which by habit keeps us quiet" and replaced silence with knowing and speaking.

NOTES

1. This chapter is a translated and expanded version of Montecino 1993.
2. Editors' Note: We understand the author's use of "the South" to refer to a specific (postcolonial) political category indicating not just Latin America but also Africa, South Asia, and, to a lesser extent, the rest of Asia. Because the author is writing from the position of "a scholar in the South," she uses the pronouns "our" and "we" rhetorically to flag that subject position.
3. See Guzmán (1990) on the fluctuation between the white and the nonwhite in *mestizo* discourse.
4. This is a Quechua concept that designates a foundling, but also an emotional state of solitude and abandonment. In Chile the term is used to talk about illegitimacy and the lack of filiation, to designate the ones who have been "domesticated," and to define the solitary and precarious.
5. Among others Camacho (2001), Cuvi and Martínez (1994), De la Parra (1996), Fuller (1993), Guzmán (1996), Guzmán and Portocarrero (1992), Hurtado (1993), and Viveros (1998).
6. Editors' Note: La Malinche refers to the indigenous translator and mistress of the sixteenth century Spanish conqueror of Mexico, Hernán Cortés.

REFERENCES

Arguedas, José María. 1978. *Deep Rivers*. Austin: University of Texas.
———. 1983. *Yawar Fiesta*. Austin: University of Texas.

Camacho, Gloria. 2001. "Relaciones de género y violencia." In *Antología de estudios de género,* edited by Gioconda Herrera. Quito: Flacso. 115–130.

Cuvi, María and Alexandra Martínez. 1994. *El muro interior. Las relaciones de género en el Ecuador de fines de siglo XX.* Quito: Ceplaes.

De la Parra, M. Antonio. 1996. "Sobre una nueva masculinidad o del padre ausente." In *Diálogos sobre el género masculino,* edited by Sonia Montecino and María Elena Acuña. Santiago: Programa Interdisciplinario de Estudios de Genero, Universidad de Chile (PIEG). 37–48.

Fuller, Norma. 1993. *Dilemas de la feminidad. Mujeres de clase media en el Perú.* Lima: Pontificia Universidad Católica.

Guzmán, Jorge. 1990. "La categoría blanco, no blanco." *Revista Tópicos.* Santiago: Ediciones Rehue, Centro Ecuménico Diego Medellín.

—— 1991. *Diferencias latinoamericanas.* Santiago: Universidad de Chile.

—— 1996. "Ejes de los femenino/masculino y de lo blanco/no blanco, en dos textos literarios." In *Diálogos Sobre el Género Masculino,* edited by Sonia Montecino and María Elena Acuña. Santiago: PIEG, Facultad de Ciencias Sociales, Unversidad de Chil. 49–62.

Guzmán, Virginia and Patricia Portocarrero. 1992. *Construyendo diferencias.* Lima: Ediciones Flora Tristán.

Hurtado, Josefina. 1993. "Mujer pentecostal y vida cotidiana." In *Huellas. Seminario mujer y antropología,* edited by Sonia Montecino and María Elena Boisier. Santiago: CEDEM. 73–86.

Kirkwood, Julieta. 1987. *Feminarios.* Edited by Sonia Montecino. Santiago: Ediciones Documentales.

Lagarde, Marcela. 1990. *Madresposas, monjas, putas, presas y locas. Estudios de los cautiverios femeninos.* Mexico D.F.: Universidad Nacional Autónoma de México.

Montecino, Sonia. 1984. *Mujeres de la tierra.* Santiago: Ediciones CEM.

——. 1991. *Madres y huachos.* Santiago: Cuarto Propio.

——. 1993. "Proposición de paradigmas para la comprensión del género en América Latina." In *Huellas. Seminario mujer y antropología,* edited by Sonia Montecino and María Elena Boissier. Santiago: CEDEM. 15–20.

——. 1995. Dimensiones simbólicas del accionar político y colectivo de las mujeres en Chile. "Una propuesta de lectura desde la construcción simbólica del género." In *Desde las orillas de la política. Género y poder en América Latina,* edited by Luna Vilanova. Barcelona: Universidad de Barcelona.

——. 1996. "Migración femenina mapuche, entre espejos y cristales." Presented at the Seminario Identidades Étnicas, organized by the Comunidad Mariknol, Temuco.

Palma, Milagros, ed. 1990. "Malinche." In *Simbólica de la femininidad. La mujer en él imaginario mítico-religioso de las sociedades indias y mestizas.* Ecuador: Colección 500 Años, Abya-Yala.

Paz, Octavio. 1989. *Sor Juana Inés de la Cruz or Deceptions of Faith 16.* Mexico City: FCE.

Toledo, Víctor. 1993. "Historia de las mujeres en Chile y la cuestión de género en la historia social." In *Huellas. Seminario mujer y antropología,* edited by Sonia Montecino and María Elena Boisier. Santiago: CEDEM. 51–64.

Vega, Imelda. 1986. *Aprismo popular: Mito, cultura e historia.* Lima: Editorial Tarea.
———. 1990. "Doña Carolina. Tradición oral, imaginario femenino, y política." In *Simbólica de la femininidad.* Quito: Colección 500 Años, Abya-Yala.
Viveros, Mara. 1998. "Quebradores y cumplidores: biografías diversas de la masculinidad." Presented at the Encuentro Regional la Equidad de Género en América Latina y el Caribe. Santiago: Flacso.

14. Local/Global ✍

A View from Geography

Altha J. Cravey

The research presented in this volume challenges us to conceive of places as uniquely constituted and produced by local inhabitants, their everyday negotiations, and their on-going struggles to shape their lives. Place is thus understood as dynamic and open, constantly recreated through social processes that are, in turn, deeply embedded in particular places. In harnessing desalambrar, a verb that is an explicit call to action, to a gendered analysis of power dynamics in specific places, institutions, and ideologies, Hurtig, Montoya, and Frazier also challenge scholars to reexamine the categories upon which our own intellectual praxis proceeds. I do so here by engaging, as a feminist geographer, in discussion with feminist anthropologists about the nature of place (and the "local"), in the hopes of devising new geographies of feminist intervention.

ETHNOGRAPHIES AND GEOGRAPHIES

Contributors to this volume open up dialogue between feminist geography and anthropology by explicitly linking questions of identity, subjectivity, and culture to everyday lived experience in specific places. These ethnographies demonstrate the power of place in shaping social identity throughout Latin America. The significance of the embeddedness of human experience comes through in the rich details of the accounts, whether the focus is on sex workers (Lamas, this volume), secondary school students (Hurtig, this volume), rural Nicaraguan women (Montoya, this volume), or female leadership (Cervone, this volume). Using gender analysis, these scholars avoid essentializing social identity or place identity by documenting the openness of both categories. Feminist geographers can benefit from the careful attention given to the specificity of local practices in these ethnographies, particularly the specificity (and interconnection) of cultural and material facets of place-based struggles, as in Klein's chapter on travesti social identities or de la Cadena's chapter on highland Peruvian marketwomen.

Feminist geographers can also benefit from the nuanced way gender is shown to order inequalities other than gender and the way this process operates in specific places in Latin America. These analyses offer a radically engaged potential—a way to identify and utilize points of leverage for social change, providing links to sites and to actors that might otherwise remain un- or underanalyzed. This method thus allows feminist geographers to reimagine and reactivate social science in the service of social activism.

What might feminist geographers offer in return for these anthropological insights and inspirations? Feminist geographers use the two highly contested and key disciplinary concepts of place and space to define each other such that place is considered "a portion of geographic space" (Johnston et al. 2000, 582).[1] Thus, on the one hand, places are settings in which social relations and identities are constituted, while, on the other hand, space is produced through social practices operating across larger geographic domains. While space has sometimes been viewed as an empty "stage" for social practices, most contemporary geographers would argue that space is as inherently dynamic as time and that social change should be understood in terms of "time–space." While this approach does not tell us much in broad terms, it is a useful technique, once we particularize its components. By conceiving of place in terms of space or more accurately, space–time, the dialectical perspective—that views places as always embedded in larger spaces—allows one to blur the boundaries of specific places and ignite new understandings of these places, from fresh vantage points. Furthermore, these insights can inflame action on multiple geographical scales—from the local all the way outward to the global—as we begin to see that daily life is never entirely confined to what may appear to be bounded geographical locations. Thus, this tactic of simultaneously imagining space and place offers a material, spatial, and metaphorical way to pursue Hurtig, Montoya, and Frazier's ambition of "tearing down the fences" (Hurtig, Montoya, and Frazier, this volume) not only in specific places, or across Latin America, but in places throughout the world.

My own research on women's lives in the maquila zones of northern Mexico illustrates this potential (Cravey 1997, 1998a,b). Exploring how women's lives were changed "on the global assembly line" and in turn, how women (and men) influenced Mexico's development trajectory allowed me to link intimate arenas with processes that I believed to be operating at multiple geographical scales. Using parallel case studies of a border town and a town dominated by a previous model of industrialization, I documented the simultaneous reorganization of social reproduction and production in northern Mexico. I found that households (and makeshift households such as single-sex dormitories) were sites of intense negotiation, while channels of public social provision were much less abundant than in the interior locations of the previous factory regime. These insights about the daily lives of maquila workers are highly specific to the situation in northern Mexico yet still may be profitably linked to insights and activism in emerging factory zones in East Asia, Guatemala, or elsewhere (Cravey unpublished ms.; Traub-Werner and Cravey, forthcoming 2002). Indeed, antisweatshop activists (such as United Students Against Sweatshops) use work such as mine to link activism in various places. Problematizing the space (and social reality) that lay beyond my case study sites allowed me to speculate more carefully about potential alliances and avenues of social change in northern Mexico.

From a geographer's perspective then, the reimagining of social science could also involve identifying potential strategic alliances that might enhance local resources and power by linking activists' concerns in disparate locations. The extended case method that I used for the Mexico research is well adapted for such purposes because of the way the *space* of the global may be deliberately linked with the *place* of the local, thus operationalizing a place/space dialectic. Incorporating and problematizing geographical and historical contexts as part of the research technique, this ethnographic method has the potential to stretch outward from micro activities to macro processes (i.e. from the local to the global) (Burawoy et al. 2000; Burawoy 1985). I believe this approach will be useful for feminist scholars and activists who want to connect the dots and link place-based resistance efforts in diverse Latin American places and beyond.

The editors and contributors of this volume have worked toward a critical understanding of the "specific, located, incarnations of national and global forces that both illustrates and makes possible the intervention of ordinary people in processes of political–economic reconfiguration and social change" (introduction, this volume, 9). As a feminist geographer I would suggest, however, that effective intervention requires us to link this kind of local, place-based awareness to the operation of those larger scale, and often global, forces. If we neglect these wider processes by treating the global as "context," rather than actually *showing* how dynamics between that realm, on the one hand, and the national and local, on the other hand, work together, we risk constructing another fence—around place—for future activists and scholars to dismantle. Thus a geographer might focus more attention on the ways that ordinary people shape worlds that extend beyond their everyday routines, even if these routines appear to be predominantly local.

In this way, place-based ethnographies can be expanded to include the geographical dialectic of local and global. This approach would thus recognize that the locally embedded lives examined here have tremendous influence on our collective global future, while these lives in turn are being shaped by what our editors refer to as "translocal processes." The relational definition of place developed in this volume would thus be extended in scope to include those actors and actions that impact many diverse locations worldwide (whether these be so-called antiglobalization activists or bureaucrats at the World Trade Organization). By employing a local/global dialectic, I do not mean to imply that "bigger is better," simply that geographically expansive processes do have an impact on many lives.

I believe the project of desalambrar can have more impact if we recognize that geographical scale is actively produced and that ordinary people create global realms as well as local ones (Smith 1992; Marston 2000). That is, I am convinced that theorizing place—as these contributions do—is a crucial first step, but one that is ultimately insufficient for the ambitious agenda of radical social change, desalambrar. Indeed, while theorizing place allows us to blur the line around local places that have constrained ethnographic projects and our understanding of these same places, we need a definition of place (and the local) that pushes our research to not only incorporate the translocal elements that shape (and are shaped by) local actors but that also foregrounds these forces as important sites of analyses. This perspective would facilitate an ethnographic method that itself incorporates techniques to link micro and macro processes. Close attention to wider arenas of social activity can provide a measure of balance for our passions and lived connections to places while showing us commonalties among diverse place-based struggles. In addition, shifting our attention to wider realms can help to illuminate key points of leverage in global systems that may not be immediately apparent from locally oriented perspectives that may sometimes constrain our vision of what is possible.

FROM THE LOCAL TO THE GLOBAL

In much the same way that anthropologists in this volume define place, geographers use analytical spatial categories, such as place, space, local, global, to describe social phenomena that are simultaneously social, material, symbolic, and discursive. Ortiz's chapter (this volume) on entrepreneurs in El Paso, Texas, lays out the global/local relationship in a way that connects anthropological insights to geographical ones. The El Paso women Marcela Torres and Carmen Rocha respond in unique ways to a U.S.–Mexico border context that has been markedly and rapidly transformed by profound and seemingly inexorable global capitalist processes. Yet their local actions, in turn, shape on-going regional and global realities. Marcela Torres, a small business owner, created a maquila directory that is sold throughout the world, while Carmen Rocha, identifying with and successfully advocating for local workers, eventually became a key player in North American politics as the New York City-based voice of southern Mexico's Zapatista Army. In a complete reversal of the popular environmentalist slogan "Think globally, act locally," the intensely local concerns of each of these women facilitate their participation in actions that impact diverse locations throughout the world.

Another way that Ortiz's work opens a dialogue with geography is through its sensitivity to geographical scale. The El Paso entrepreneurs intervene in local, regional, national, and global affairs, and do so in different ways, depending on the choices they make. At the same time, Ortiz recognizes that these distinct geographical scales are products of social history and not preordained or fixed in some permanent relationship or hierarchy. Thus, the national realms of both Mexico and the United States have been shaped and produced by complex social histories involving conflicts and negotiations within and between these countries. In recent decades, economic integration has bound the two national spaces in an ever-tighter embrace and, in 1994, the North Atlantic Free Trade Agreement (NAFTA) institutionalized and formalized a process that had long been visible in the borderlands region. The point is that these nested geographical scales are actively produced. Ortiz's work, while grounded in the everyday lives of individuals, illuminates the way their lives and actions intersect various spatial scales, disrupting or perhaps reinforcing the connections between and among them.

How might a geographer extend other ethnographies in this volume to the global realm, or at least to causal relations operating at wider geographical scales? Frazier's examination (this volume) of a human rights project in northern Chile makes many of these wider linkages explicit. Members of the ex-prisoners association in Iquique understand their situation and their ultimate goals as part of something much more geographically extensive—the nation-state. They make demands that engage the nation-state while in no way seeking to "create an autonomous realm" (Frazier, this volume, 110). Furthermore, Frazier acknowledges the active role of transnational human rights discourse and alliances of international solidarity in framing the relationship that Iquiqueños pursue at the national level.

Moving from the scale of the body to transnational discourses, Frazier accomplishes a lot in a few pages, yet a feminist geographer might choose to make these wider linkages even more explicit. The Iquique experience clearly has implications at national and global scales yet such speculation requires asking questions about these spatially extensive relationships. For instance, a geographer might ask whether the exhumation of mass graves in Iquique has contributed to transnational human rights discourse? Has the local struggle had an impact on national and international interpretations of the Pinochet years? In what ways does the metaphor of "health as politics" contribute to or undermine the coercive hegemonic project of neoliberalism that is also widely known as the "Washington consensus" (ADD PAGE #S)? These questions are obviously beyond the scope of Frazier's concern about how human rights activists in Iquique transformed their assigned place in relation to the nation and national history. However, this global/local line of investigation could open avenues for alliance with activists who had similar concerns to those in Iquique. These potential alliances might include others working on intensely local and national projects of human rights (e.g., disappearances and their consequences during Argentina's military regime) as well as activists who focus more attention on global discourse or "alternative" (i.e., social justice oriented) avenues of globalization.

Feminists—in fields such as anthropology and geography—suggest that places are best understood by thinking about particular sets of intersecting social relationships. By colliding (or not) in a particular location, these social relationships produce a unique local mix. Distinct places—homes, villages, and regions such as Latin America—are thus theorized as being produced by the intersecting social relationships that meet in a particular time and place. The attention to specificity herein provides new ways to harness our own transgressive potential in pursuit of locally based action and locally based alliances. It is in local places after all, that we can most readily find openings in cultural political economic systems that all too frequently appear to be hegemonic (Gibson-Graham 1997). In closing, let me endorse the place-based passions that Eudora Welty described as the "fire that never goes out" (Welty 1998, 760).[2] Let us fan these flames, tear down more fences, stoke the inferno, and meanwhile, incite our neighbors as well as those far from home to join us.

NOTES

I would like to thank the editors of this volume for provoking dialogue that deliberately transcends disciplinary and academic boundaries.

1. Massey 1994; Jones, Nast, and Roberts 1997; Women and Geography Study Group 1997.
2. Welty wrote, "A place that was ever lived in is like a fire than never goes out" (1998, 760).

REFERENCES

Burawoy, Michael. 1985. *The Politics of Production: Factory Regimes under Capitalism and Socialism.* London: Verso.

Burawoy, Michael, Joseph, A. Blum, Sheba George, Zsuzsu Gille, Teresa Gowan, Lynne Haney, Maren Klawiter, Stephen H. Lopez, Sean O Riain, and Millie Thayer. 2000. *Global Ethnography: Forces, Connections, and Imaginations in a Postmodern World.* Berkeley: University of California Press.

Cravey, Altha, J. 1997. "The Politics of Reproduction: Households in the Mexican Industrial Transition," *Economic Geography* 73(2): 166–186. April.

———. 1998a. *Women and Work in Mexico's Maquiladoras.* Lanham, M.D.: Rowman and Littlefield, Inc.

———. 1998b. "Cowboys and Dinosaurs: Mexican Labor Unionism and the State." In *Organizing the Landscape: Geographical Perspectives on Labor Unionism,* edited by Andrew Herod. Minneapolis: University of Minnesota Press. 75–98.

———. "Toothless Tigers and Mouldered Miracles: Geography and Global Gender Contracts in the NICs." Unpublished manuscript.

Gibson-Graham, J. K. 1997. *The End of Capitalism as we knew It: A Feminist Critique of Political Economy.* London: Blackwell.

Johnston, R. J., Derek Gregory, Geraldine Pratt, and Michael Watts. 2000. *The Dictionary of Human Geography.* 4th Edition. London: Blackwell.

Jones, John Paul, Heidi J. Nast, and Susan M. Roberts, eds. 1997. *Thresholds in Feminist Geography: Difference, Methodology, Representation.* Lanham, Maryland: Rowman and Littlefield, Inc.

Marston, Sallie. 2000. "The Social Construction of Scale." *Progress in Human Geography* 24(2): 219–242.

Massey, Doreen. 1994. "A Global Sense of Place." In *Space, Place and Gender.* Cambridge: Polity Press, 146–156.

Massey, Doreen and Pat Jess, eds. 1995. *A Place in the World?* Oxford: Oxford University Press and Open University.

Smith, Neil. 1992. "Contours of a Spatialized Politics: Homeless Vehicles and the Production of Geographical Scale." *Social Text* 33: 54–81.

Traub-Werner, Marion and Altha J. Cravey. 2002. "Spatiality, Sweatshops, and Solidarity in Guatemala." *Social and Cultural Geography* (forthcoming).

Welty, Eudora. 1998 (originally published 1944). "Some Notes on River Country." In *Eudora Welty: Stories, Essays, and Memoir.* New York: The Library of America. 760–772.

Women and Geography Study Group of the Institute of British Geographers. 1997. *Feminist Geographies: Explorations in Diversity and Difference.* Essex: Longman.

Postscript ∽

Gender in Place and Culture

June Nash

The current tendency to dis-place place in asserting that global processes force everything into flux threatens to undermine one of the principal coordinates of ethnographic discourse. The editors and contributors to this volume address how social practices continue to relate gender to place and how these are transformed in the course of political struggle. They do this by wedding the material and ideological perspectives of gendered ethnographic analysis, building on the specificity of local practices and meanings and grounded in participant observation of all aspects of life.

The theme of this volume is the impact of fencing in and of contesting boundaries defined by gendering space. To desalambrar, or to take down the wire fence that surrounds one in a space or, metaphorically a category, opens new spaces and possibilities. Many of the papers refer to the excruciating moments when women and/or men become conscious of restrictions on their bodily presence in a place from which they have been restricted or to which they have been consigned. The violence to person put in place by such restrictions is revealed in the pain as well as pleasure in realizing freedom from such strictures.

REFLECTING ON PLACE AND GENDER

This collection of articles promotes self-introspection as we explore those moments of self-realization in other places and other times. My first experience of being corralled in a gendered space came during World War II when I was sixteen years old. I worked on what was euphemistically called a "Victory Shift" of the local Sylvania electrical plant that allowed high school students to participate in an electronics factory assembling radar units for battleships. The huge assembly room contained over 500 women facing each other on an assembly line across a table with two moving belts carrying the assembled units with the parts to be mounted to each worker's station. The hum of hundreds of conversations across the lines punctuated the daily routines of repetitive motions in one-minute cycles. The songs piped through the loudspeaker promoted the passive female role, presumed to be just sitting and waiting for the return of the troops: "You'd be so nice to come home to," "Missed the Saturday dance." The jobs involving physical motion were done by men; boys not yet of draft age or men too old to be recruited stacked the material to be assembled at the head of the lines and packed the finished goods after they were tested. Managerial and engineering jobs were held exclusively by men.

Looking back at that setting with a gendered perspective I could appreciate the cloistered atmosphere in which every movement was dictated and controlled by an all-male management in terms that conformed to their notion of the passive, almost inert, females. In the Taylorite regime of work, women's jobs were dormant, involving microscopic movements limited to a one-minute cycle repeated endlessly. I still remember my job: Arching three wires extending from the conical casing and terminating them with a loop, all within a one-minute cycle. Managers appeared occasionally to survey the room of 500 females, and gave orders to the "line girls"; women who knew every job on the line and who sat in for girls who had to take a break. We received twelve dollars a week for five days of four-hour work shifts.

One day, without advising the workers what the game plan was, a half-dozen men entered the room and ordered the "line boys" to set up a stretch of assembly line in the corner of the room. Rumors ran up and down the line—it was a new form of speedup some guessed; others proposed they were introducing a new experimental product. Twenty women were chosen to form the new line. The first day's work was reportedly normal but on the second day the women showed up to find a six-inch-high plank dividing the two lines of women facing each other down the center of the assembly table. No explanations were given, but rumors were rampant as the women found an additional dividing plank in place each day on their arrival. When the divider rose to their shoulder level, they realized that the game plan was to limit conversation across the table. Next day when the men arrived in their morning inspection, the experimental group broke into song, "Don't fence me in!" They were joined by all 500 women in the cavernous assembly room. Seeing that they were defeated, the men retreated, and the next day, the experimental line was dismantled.

The corralling of women that was accepted until it exceeded the norms of that 1940s workplace would seem absurd today were it not for the Taliban experiment in controlling women's movements to the point that it threatened not only their livelihood but their lives. Documentation of the suffering and even starvation of women who could not go out to work or gather food, even when they had no male household members, was disclosed in e-mail bulletins throughout 2001 and after 9/11 in daily news reports from Afghanistan in the war on terrorism.

GENDER IN THE GLOBALIZATION PROCESS

Reports of these aggressions against women in Afghanistan and other Muslim nations evoke a response to globalization processes that I have been looking at in Latin America (Nash 2001, n.d.). The increase in male violence against women is particularly visible in areas of ethnic resurgence among indigenous peoples. The dual threat of men losing control over the subsistence economy and over women as they are forced to seek alternative sources of income outside of household production seems to trigger the rage men feel and direct against women. This was acutely apparent in Chiapas, Mexico after the Zapatista uprising in 1994. Women's participation in the uprising and in the autonomy movement that issued from it gained international acclaim, with *comandantes* Ramona, Ana María, and Trinidad, who often outshone their male comrades David and Tacho at their international conventions. But in the local scene, the violence against women escalated. The Acteal Massacre of 1997 in which 36 women and children, along with nine men, were killed by neighbors and relatives in the indigenous township of Chenalhó made front-page news internationally. Eyewitnesses say that the youths, recruited into the paramilitary group called the "Red Masks," ranged in age from 14 to early 20s, and were supervised by an army sergeant and retired general. Some reported that they seemed crazed with drugs, as they sportively ripped open the bellies of four pregnant women and tossed the fetuses from one bayonet to another.

The massacre is an extreme example of gender antagonism that is an offshoot of globalization processes that favor women in low-paid assembly work and that undercuts the subsistence cultivation in which men traditionally engage. In the case of the Acteal massacre, the youths who participated in the attack did not see a future for themselves in small-plot cultivation as land grants had been rescinded by the neoliberal government and surplus crops were no longer protected in global markets. Indigenous people are fighting for survival in the global arena, but gender antagonism threatens the solidarity of the movements that have evolved in the hemisphere, especially since the 1992 counter celebrations in recognition of 500 years of colonization and marginalization. Women are among the top command of the Zapatista movement, joining in the armed struggle that broke out on New Year's eve in 1994 and remaining in the vanguard of opposition to the militarization of the region that followed. But for male rank-and-file members the engagement of women in the Zapatista mobilization constitutes a threat; one man even killed his wife when she insisted on going to meetings.

GENDER DISTINCTIONS IN ETHNIC RESURGENCE

In these ethnic resurgent movements women counter the tendency of men to exalt all traditions of the past, criticizing the alcoholism and domestic violence that was excused as ritual. Women are more likely than men to reject the power of *caciques*, the native leaders who are mediators in the exploitation of the labor and resources, as they challenge the patriarchal power structures that ensured men's control over women. Their challenge is becoming more acute as the restraints on domestic abuse that once rationalized the power of the patriarch in the nineteenth century (cf. Alonso, this volume) succumb to the contradictions facing the family in the neoliberal economies.

Indigenous women are less subject to academic rubrics or to the ideal types applied to Latin American women as saintly mothers or streetwalkers than are *mestizas* or whites. In indigenous communities the power of women as progenitors of the human race is part of a mythological and metaphorical conceptualization that predates the conquest (Nash 1997). Marysa Navarro (this volume) faults the analysts who have found in marianismo, or sacred mother, and *machismo*, or sexually charged male, a compatible model of gender relation, for their lack of empirical knowledge of women in real social contexts of either *indigenas* or *mestizas*. Like the Cartesian dichotomies of "man is to culture as woman is to nature" that so influenced Simone de Beauvoir's thought in *The Second Sex*, these conceptions have to be destroyed in order to evaluate empirical conditions in the world.

GENDER AGGRESSION IN THE TRANSITION TO EQUALITY

Patriarchy is linked with power, but often in ways that naturalize the place and temporal dimensions of categories of people occupying inferior or subordinate places. These inferences drawn at subconscious levels are exposed in moments of desalambrar. Among such cases is the transition from a military regime to democracy in Chile. Lessie Jo Frazier (this volume) captures the dynamics of the attempt by men to regain control over women in her analysis of civil society activists who challenged the "hyper-masculinized" military regime of Pinochet when he still wielded power. Their denunciation of the violators of human rights culminated with the discovery of the cemeteries of disappeared people and their repossession of these as sites for political struggle. In the process, they encountered a tendency on the part of the government to engender human rights activism and mental health resources as feminine. Frazier shows how the recuperation of democracy became depoliticized as state violence was translated into individual experiences of mental breakdown.

Adolescent males face an uncertain future as women gain greater access to the jobs that have come with globalization of the economy. Women's growing involvement in wage labor and the cash economy defies male categories for defining women's place just as it unmasks the bread-winning role of males that underwrites patriarchy. These conditions, added to the disjuncture in gender opportunities in the labor market, generate passions that cannot be conflated in class analysis. The "transgressions" of women in places dominated by men—the streets, public places in general—fertilize the gender antagonism that erupts in unpredictable ways. Whereas Nicaraguan village women have negotiated more open spaces in their own terms (Montoya, this volume), the phenomenon of transgressions by sex workers on the streets of Mexico City at night presents a far more dangerous and, for some women, more challenging picture (Lamas, this volume). The male categories that define women as decent or not decent when they enter these restricted areas still dominate the discourse on sexuality. The entry of travestis into such domains complicates the picture, as the male purchaser of such street players may vent his self-hatred in violence against those who evoke homoerotic behavior. In these highly charged sexual settings, aggravated by the AIDS epidemic, the presence of nongovernmental support groups can become crucial in politicizing those who navigate in those streams (see especially Klein, this volume). The innovative ways in which agents of these transnational organizations address the problems of discrimination and hate crimes are providing a model for intervention that might well serve in other settings.

Feminist teachers and political leaders are trying to air these models in order to channel the hostility that arises in these contexts. One such step in the direction of publicizing and gaining acceptance for new practices related to women's changing roles in society are the public discussions that are now admitted into school debates, as in Venezuela (Hurtig, this volume).

When the discontent of adolescent males with their marginalization in the world economy is ignored, it may find expression in the recruitment of youths into militant religious and paramilitary units that allow them to discharge the rage they experience onto women.

In what Lessie Jo Frazier calls "the gendered processes through which local actors engage with and attempt to negotiate dominant boundaries of place" (this volume, page 93), the structures of social practice of desalambrar can be restructured in response to the new social reality. These are the moments that some are calling epiphanies in the sense of the discovery or sudden manifestation of the essence or meaning of something. It may come with the assertion of a collective proprietorship of the political landscape in Chile, as Frazier asserts, or with Nicaraguan women taking back the village streets. The sense of discovery, or epiphany, is the awareness of rights that one had been unaware of and the empowerment in its rediscovery. This was what struck me when the line workers rejected the control instituted by the managers in the electronics plant; much later I experienced it when I joined a "Take back the night" movement organized by the women of Pittsfield, Massachusetts when I was doing research in this industrial town. These epiphanies are the sparks that ignite rebellion but do not always result in change in the power structure. We never did get union representation throughout the war when we worked for 50 cents an hour, but young women are venturing out in Pittsfield unescorted in the night hours.

Those moments of insight are heightened when social subjects confront double barriers of gender and ethnic restriction. Ecuadorian indigenous women experience the catalyzing responses to ethnic mobilization by questioning the gender barriers left in place. Emma Cervone (this volume) shows how the unleashing of gender discrimination operates in the context of intense ethnic mobilization in Ecuador. Indigenous women were not able to take on leadership roles in the new government settings brought about by access to regional power because they lacked the education and Spanish language ability that men had acquired under the old regime. However, women exposed the specious claims to equality as they began to negotiate a space in the polity in their own right. Much as in Chiapas, women had to unmask the violence and abuse latent in indigenous society in order to advance their own claims for equality in the new ethnic accord. As indigenous members of a racist society, women are still stigmatized by men when they make charges of sexism in indigenous meetings. Clearly gender claims have not achieved legitimacy within indigenous movements to the same degree that they have in mainstream society, yet in breaking the barriers to their participation in democratic processes women are changing the structures of both racist and sexist domination (Cervone, this volume).

CLASS AND SEXUAL AUTONOMY

Postmodern analysts deliberately eschew the economic basis for gender activism, but it is clear that the drive for sexual autonomy is accelerated by women gaining independent sources of income. This often occurs in gendered settings where women control their own spaces: Market women in the Andes are renowned for their forthright demeanor, their independence of male providers, and their ability to command sexual attentions in their own terms. When I worked in Bolivia I was impressed by the way that market women, particularly meat sellers, commanded a vibrant local economy in Oruro that enabled them to gain political power by backing politicians of their choice. They also had a myriad of children by a variety of men who were burden bearers and performed other functions in the market, but the last thing they wanted was to have dependent husbands. Marisol de la Cadena (this volume) shows how market women of Cuzco in the decades from 1930s to 1960s rejected the categorization of women in terms of middle-class morality. Race and class intersect to produce complex signifiers of the social position of these women, and it is in the market place that these intersecting vectors of change have the most impact.

Victor Ortiz's analysis of industrialization at the Mexican–U.S. border shows the interrelation of class and gender in the global reconfiguration occurring at the border with the United States. His comparison of a female labor leader and a female business owner who promoted the shift from light manufacturing to professional and technical services allows him to explore the transformations occurring as women gain personal and political agency in a region that is often taken as a global disaster area. Though coming from different sides of the manager–worker negotiation table, both women exemplify the attempt to maintain jobs by integrating the concerns of women as producers, consumers, and intermediaries. The greater autonomy that women gain in wage work in the *maquila* manufacturing sector is still modified by blocks to translating economic success into political and social status. The rank-and-file women in assembly plants are still subject to male overseers who manage their motions in accord with sexist definitions of women's roles in production.

DECONSTRUCTING BARRIERS AND CATEGORIES

The concept of desalambrar sharpens our awareness of how categorical correlations of people with places are constructed in ways that conduce to critique and controversy. This in itself does not succeed in doing away with hierarchy, but it can provide insights that enable the subordinate to undermine the practices that reinforce structures of domination and subordination. There remains the challenge of applying the insights gained to cultivate the practice of egalitarian relations in all institutional contexts.

Gendering of place remains an important signifier of status, but in the changing global context the meanings are shifting along with the behaviors that correspond to these changes. Women now play a greater role in deconstructing the old shibboleths about propriety, but as barriers fall in the centers of modernization and change the backlash from men in societies marginalized by globalization can still be felt on a larger scale than ever before. Freedom from the ideological and material fencing in cannot be achieved simply by cutting away the behaviors of sexism; it will ultimately depend on gaining economic and social structures that will allow women to escape patriarchal controls.

REFERENCES

Nash, June. 1997. "Gendered Deities and the Survival of Culture." *Journal of the History of Religion* 36(4): 333–356.
———. 2001. *Mayan Visions: The Quest for Autonomy in an Age of Globalization.* New York and London: Routledge.

Notes on Contributors ⌇

ANA MARIA ALONSO is Associate Professor of Anthropology at the University of Arizona–Tucson. She is the author of *Thread of Blood: Colonialism, Revolution and Gender on Mexico's Northern Frontier* as well as articles on social memory, popular resistance, nationalism and ethnicity, AIDS and sexuality, agrarian reform, and the Mexican Revolution.

RUTH BEHAR is the author of *Translated Woman: Crossing the Border with Esperanza's Story* and *The Vulnerable Observer: Anthropology that Breaks Your Heart.* She is the editor of *Bridges to Cuba* and co-editor of *Women Writing Culture.* She is Professor of Anthropology at the University of Michigan.

EMMA CERVONE is Assistant Professor of Anthropology at Southern Illinois University. She has published articles on ethnicity and indigenous movements in *Anthropos* and the *Journal of Latin American Anthropology* and is completing a book that explores the politics of ethnicity in the contemporary Ecuadorian indigenous movement.

ALTHA J. CRAVEY began to interrogate geographies of gender, work, and globalization while working as a construction electrician. Now Associate Professor of Geography at the University of North Carolina, she is the author of *Women and Work in Mexico's Maquiladoras.*

MARISOL DE LA CADENA is Associate Professor of Anthropology at the University of North Carolina–Chapel Hill and a member of the Instituto de Estudios Peruanos in Lima, Peru. Her book, *Indigenous Mestizos: The Politics of Race and Culture in Cuzco, Peru (1919–1991),* is a historical ethnography of intellectual–political dialogues among socially diverse groups on issues of race, gender, class, and geography.

LESSIE JO FRAZIER is an anthropologist and historian working on issues of gender, labor, memory, human rights, nation-state formation, social movements, and activism in Chile and Mexico. Her works in progress include

Salt in the Sand: Memory and Violence in Chile and the co-edited volume *Love-in, Love-out: Gender and Sexuality in a Global 1968.*

JANISE HURTIG (Center for Research on Women and Gender, University of Illinois at Chicago) does conventional and participatory research on gender, literacy, schooling, and social change in Venezuela and Chicago. Her book *Coming of Age in Times of Crisis: Schooling and Patriarchy in a Venezuelan Town* is forthcoming (Palgrave).

CHARLES H. KLEIN is a Health Program Planner at the HIV Prevention Section of the San Francisco Department of Public Health. He received his Ph.D. in Anthropology from the University of Michigan in 1996. His current research focuses on queer sexuality and sexual health movements in Brazil and the United States.

MARTA LAMAS studied anthropology at the Escuela Nacional de Antropología e Historia (ENAH) in Mexico. She has been active in the feminist movement for the last 30 years. She is the editor of the journal *Debate feminista.*

BARRY J. LYONS, an anthropologist at Wayne State University, studies agrarian conflict, religion, ethnicity, and the sociolinguistics of exchange in highland Ecuador. He is currently completing a book on the hacienda and its legacy.

SONIA MONTECINO, anthropologist and fiction writer, received her doctorate from the Universidad de Barcelona. She is Professor of Anthropology and Director of the Centro Interdisciplinario de Estudios de Género of the Universidad de Chile.

ROSARIO MONTOYA received her Ph.D. in Anthropology and History from the University of Michigan. She writes on popular culture and religion, class, gender and sexuality, and nation-state formation in Nicaragua. She is completing a book entitled *Ambivalent Revolutionaries: Exemplarity and Contradiction in a Sandinista Model Village, Nicaragua, 1979–1999.*

JUNE NASH organized with Helen Safa the first conference on feminine perspectives in 1974, resulting in the publication of the book, *Sex and Class in Latin America.* She has carried out anthropological research on community, family and gender roles and the impact of globalization processes in Mexico, Bolivia, and the United States. Her most recent book is *Mayan Visions: The Quest for Autonomy in an Age of Globalization.*

MARYSA NAVARRO is Charles Collis Professor of History at Dartmouth College. Her writings range from works on social movements, to studies of right-wing thought and women in the labor movement in Argentina, to a biography of Eva Perón, including, "The Personal Is Political: Las madres de la Plaza de Mayo," "The Construction of Latin American Feminist Identity," and the collaborative essay, "Feminisms in Latin America."

VICTOR ORTIZ teaches at Northeastern Illinois University and is Coordinator of its Mexican/Caribbean Studies Program. He studies the role of ethnicity in the new global economy. His book, *El Paso: A Frontier from Space to Hyperspace* (forthcoming, University of Minnesota Press) examines the socio-political impacts of the economic integration of Mexico and the United States in El Paso, Texas.

SUSAN J. PAULSON spent 12 years in South America developing graduate programs and doing field research. Her publications include *Social Inequality and Environmental Degradation in Latin America* (Abya Yala 1998) and the collection *Gender Theories and Practices: A Dialectical Conversation* (Poligraf 1997), both in Spanish. She currently teaches at Miami University.

Index ᕲ

Printed in the United States
33739LVS00004B/1-45